In the autumn of 1971, Kathryn Hulme, her dearest friend Lou — on whom her famous best seller, *The Nun's Story,* was based — and their friend Juliet made a bone-shaking safari of twenty-five hundred miles in a Land Rover with an open hatch, photographing the animals and the natives, and absorbing the impact of the great national parks in Kenya and Tanzania. Lou had lived and nursed in the Congo for seven years; Juliet, the horticulturist, was making her third safari in East Africa; but to Kate Hulme, the novice, everything was fresh, seen for the first time, and to the writing of her chronicle she brings the intensity and vivid dedication that so illuminates her earlier books.

From the moment of their arrival in Nairobi, the trio was in the expert care of an English escort-driver named Luke, a war veteran fluent in Swahili, with an extraordinary, sensitive knowledge of the habitat and character of the wildlife they had come to observe. With him, they were blessed with good luck in where they were to go and what they were to see. They saw more than a million flamingos feeding or at rest on their fabulous soda lake, lions sleeping in the trees at Lake Manyara, zebra stallions who came to bathe within fifteen yards of where Kathryn Hulme was standing, a two-thousand-pound mother rhino who couldn't quite make up her mind to charge their Rover, a fiercely proud Masai warrior who unexpectedly opened his cape and drew out a gift from a small leather bag. They journeyed from Nairobi to Samburu, where they came face to face with their first lion, across the desert to Marsabit, the land of the biggest elephant on earth. They climbed Mount Kenya, explored Olduvai Gorge, site of one of the world's greatest fossil lodes, and trekked northward again through Arusha and the Amboseli Game Reserve.

All of this was accomplished with two of the travelers in their seventies and the third a grandmother in her sixties: "as if once a woman becomes eligible for Social Security,

BOOKS BY KATHRYN HULME

ARAB INTERLUDE

DESERT NIGHT

WE LIVED AS CHILDREN

THE WILD PLACE
(*Atlantic Nonfiction Prize Award, 1953*)

THE NUN'S STORY

ANNIE'S CAPTAIN

UNDISCOVERED COUNTRY

LOOK A LION IN THE EYE
On Safari Through Africa

she is done for as far as rough travel is concerned!" Under Luke's guidance, they jounced along for days across the great plains. And in the evenings, drinks in hand, under canvas or at a guesthouse as spectacular as the one called The Ark, it was a time for summing-up and for sharing their experiences of an Africa both majestical and mysterious.

In *Look a Lion in the Eye,* Kathryn Hulme shares those experiences with us, too, writing about them in a prose that is lovingly crafted, rich in memorable detail and filled with warmth. Her evocation of Africa, her concern for the wildlife, her wonder and humor all form a glowing whole that rekindles a continent for all those who have been there in the flesh — or only in their dreams.

LOOK A LION
IN THE EYE

LOOK A LION
IN THE EYE

On Safari Through Africa

BY KATHRYN HULME

AN ATLANTIC MONTHLY PRESS BOOK

Little, Brown and Company — Boston – Toronto

FIRST EDITION

T 05/74

Library of Congress Cataloging in Publication Data

Hulme, Kathryn Cavarly, 1900-
    Look a lion in the eye; a safari memoir.

    "An Atlantic Monthly Press book."
    1. Zoology--Africa, East. 2. Africa, East--
Description and travel. I. Title.
QL337.E25H84   599'.09'676   73-19519
ISBN 0-316-38140-3

ATLANTIC—LITTLE, BROWN BOOKS

ARE PUBLISHED BY

LITTLE, BROWN AND COMPANY

IN ASSOCIATION WITH

THE ATLANTIC MONTHLY PRESS

Published simultaneously in Canada by
Little, Brown & Company (Canada) Limited

PRINTED IN THE UNITED STATES OF AMERICA

For
JULIET
RICE
WICHMAN

Foreword

IN THE AUTUMN OF 1971, in a party of four including the escort-driver, I traveled for one month and some twenty-five hundred miles in a bone-shaking Land Rover through Kenya and northern Tanzania, "viewing game" as they say, in the national parks of East Africa. The impact of Africa on the newcomer (this was my first safari) is what this book is about. By impact, I don't mean the blows sustained when the Land Rover bucked over unmapped game trails, but the striking force with which Africa impresses on first view — a lasting impression that doesn't fade away like body bruises.

Until I went on one, "safari" was something of a mystery word for me, although I knew it derived from the Arabic for journey. What was it like to "go on safari"? What was the day-to-day experiencing? Just *looking* at wildlife, or was there something more? This book tries to answer such queries, in a sequence of daily revelations just as they happened to me.

There are of course many types of safaris of varying durations (and prices), from the most usual twenty-one-day tours that take in limited areas to two-month safaris

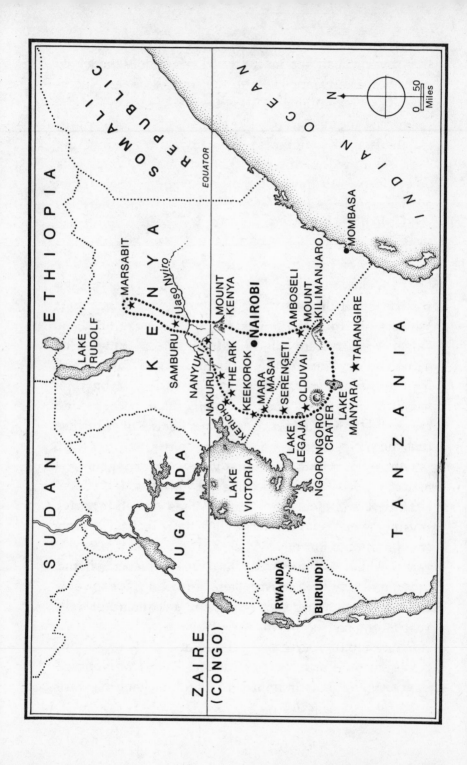

that cover, via air and motor travel, practically the whole of Africa, from Cape Horn to the sources of the Nile in Uganda's "Mountains of the Moon." There are safaris for special-interest groups like the Audubon Society bird watchers, who sometimes go out in parties numbering over a hundred, or the more usual average groups of a dozen to twenty tourists who go out under their travel agency's aegis in zebra-striped minibuses, and there are individually planned safaris, like ours, which are never booked less than two nights in any one place and allow plenty of time to see everything, even twice, if the light for photography is no good on first viewing. This is deluxe travel, topped for comfort and privacy only by the "tented safari" — the most expensive of all — which sets up camp in choice spots in the middle of game country and requires a caravan of trucks and at least a dozen servants from cook to barboy to operate. Hemingway, in the stories that came out of his big-game hunting days, made the "tented safari" a familiar setting to armchair travelers.

Our itinerary had been arranged nearly a year in advance, especially the African end of it, necessary in a land of small hotels if we were to have bed and shower waiting at each day's end. Our route described a regal sweep through twelve of East Africa's faunal national parks, which had been set aside, for the protection and preservation of wildlife, originally by the European colonials of pre-Independence days and then continuously maintained by the new young black republics as they emerged and took over in the sixties.

We started north from Nairobi to Aberdare National Park and The Ark, then through Samburu Isiolo Reserve to Marsabit on the Northern Frontier. Here we turned south, doubled back on our tracks to Mount Kenya, which we climbed; then we went west and south through Great Rift Valley country to Lake Nakuru and Mara Masai, crossed the border into Tanzania for Serengeti, Ngorongoro, Olduvai, Lake Manyara and Tarangire, then northward again through Arusha and Amboseli Game Reserve to Nairobi.

On the trip map, the dots enclosing the areas we covered describe an elongated loop roughly resembling a war club dropped across the equator, with handle to the north and knob end to the south of it. This war club analogy, completely fortuitous, could nevertheless serve as the symbol of the kind of battling determination that characterized the makeup of our party. We were three women of Kauai (Hawaii), all old friends and all old enough to be grandmothers — one sexagenarian and two septuagenarians, to be exact. I bring in these vital statistics only because some friends suggested that our age and sex were the more remarkable aspects of our safari — as if once a woman becomes certifiable for Social Security, she is done for as far as rough travel is concerned! The truly remarkable fact of the matter was that three mature women with minds of their own and three different cameras all competing for the best light could travel together for a month, boxed up daily in a Land Rover for hours at a time, and come out of that safari experience not only still on speaking terms but without once having wished to kill each other.

Somewhere I have read that more than a quarter-million tourists from other continents visit the African game reserves each year. You encounter these fellow travelers only after-hours, so to speak, in the cocktail lounges and dining rooms of the game lodges. They are an international crowd, predominantly British, Italian, German and American, representing every category of human from professor to playboy, and bearing with them just about every imaginable viewing aim or hope, as you can hear from their excited table talk.

In voices of wonder they speak of everyday sights like the wildebeest herds moving single file across the grasslands followed by their ubiquitous predators — the hyenas or the wild dogs or the swift-running cheetah, who has been clocked at sixty miles an hour; elephants blocking the road and bringing all vehicular traffic to a quick and silent halt; a Serengeti sunset resembling an atomic bomb explosion, where there are horizon clouds that light up; and above all, the sight of sights, the unique and unparalleled wonder of *Panthera leo*, the lion — sprawled in siesta, gnawing a kill, or simply sitting on top of a termite mound watching the world go by with great yellow eyes that can spot movement more than a mile away across flat plains.

I never met anyone on safari who had ever had enough of the African lion. His name in Swahili is *simba*, a word everyone learns on the first sight of him and retains long after the safari ends — a magical incantation that brings him to instant life in one's memory, all three hundred and fifty pounds of him wrapped in his royal mystery and shaggy magnificence.

I emerged from the writing of this book indebted to many people who gave me, from their specialized fields, invaluable help, criticism and advice. Space limitations preclude my acknowledging them all, but I cannot close without naming with grateful thanks the key contributors to whatever accuracy and goodness my book may have. Don Turner's experienced company in Nairobi — the East African Ornithological Safaris, Ltd., who planned our itinerary and took care of us throughout Africa — spared me endless research by sending me after I got home many essential information booklets and maps that I had neglected to pick up during my travels. Solita Solano, my literary guardian angel since I began writing, received my chapters as I completed them and from her studio home outside Paris sent them back to me instructively decorated with the marks of her firm editorial blue pencil, which cut, transposed and rearranged my material with the genius touch for which she is famous. And finally, from friends living here on Kauai came the kind of professional help one would never expect to find on a mid-Pacific island. William Shroder, a veteran photographer whose work adorns museum walls, read my manuscript for any photography errors an amateur might have committed in print and he also helped me to select from a great collection of my animal shots the best of the lot for reproduction in this book. Artist Ida Faye Dawson designed the trip map, and her sister, Margaret L. Faye, took on the drudgery of typing the manuscript twice straight through — once for Solita's editing, and once again after it, for the final copy.

K. Hulme

Contents

	FOREWORD	ix
I	GOING TO AFRICA	1
II	THE THEATRICAL ARK	11
III	NORTH TO SAMBURU GAME RESERVE	25
IV	NATIVE DANCES	43
V	BUFFALO SPRINGS AND MORE HONEYMOONERS	53
VI	KENYA'S NORTHERN FRONTIER PROVINCE	65
VII	MARSABIT — WATERING OF THE CATTLE	79
VIII	BOTANIZING ON MOUNT KENYA	91
IX	NAKURU AND THE FLAMINGOS	105
X	"ANY TIME IS TEA TIME"	119
XI	THE ADORABLE HYENA OF MARA MASAI	133
XII	WE WERE NOT "BIRDERS"	151
XIII	"WHO WERE THESE MEN?"	171
XIV	NGORONGORO — A WORLD APART	185
XV	A RARE FINALE	203

I

GOING TO AFRICA

Nairobi, capital of kenya and the customary starting point for an East African safari, startled me by looking from aloft like a city. As our plane circled in for a landing, I caught glimpses of distant skyscrapers and felt a momentary disappointment at the sudden sophistication of my first view of Africa. Not long ago, lions were occasionally seen roaming Nairobi's streets, and I had imagined it a thatch-roofed village.

We had flown to Africa from the East — from Honolulu via Hong Kong to Bombay, thence over the Indian Ocean to the East African coast somewhere in Somalia, and finally southward to Kenya and its capital city, Nairobi. Here, our small party received its sole male member, the escort-driver, waiting at the airport to pick us up.

To introduce the disparate characters of our private safari, I begin with Juliet, who began it. Nearly a year before we took off, she had come to our house one day, on Kauai, where we all lived, and invited Lou and me to be her guests on her next African safari, which would be her fourth. That she had survived three such rugged journeys, two with grown sons, their wives and young children taken along in combinations selected for compatibility, indicates the chief feature of her inheritance — a constitution of iron. Juliet was a descendant of New England missionaries, a *kamaaina* of Kauai, which means island-born and therefore quite special in our small island society, a powerfully proportioned matriarch with the height and bearing of an early Hawaiian queen. She had been having a love affair with Africa ever since her first safari in 1967.

Lou was a totally different personality. Belgian-born, long a naturalized American citizen, she had collaborated with me in the writing of *The Nun's Story* which, in the late fifties, astounded us both by making the top of the best-seller list and staying on that list for a year. The biography was in fact the story of her seventeen years in a convent from which she had "come out" toward the end of World War II, when I first met her. Seven of her convent years had been spent in the Belgian Congo as a nursing nun, so in one sense our safari was to be for her a sort of going home, although to a part of Africa she had never known as a nun. Yet her years of working with and for the Negro race had given a depth to her understanding of it that the rest of us lacked. As her long-unused Swahili freshened in her memory and she began to follow what the black hotel servants were saying about their white charges, she often smiled to herself, a secret participant in their comical summings-up.

In this trio of elderly globe-trotters, I was the novice who had never been in equatorial Africa and had simply no idea of what that miscalled "Dark Continent" does to one after intimate experiencing of its stones and bones and its endless savannas populated by the great mam-

4

mals one meets at close range with no zoo bars between.

Our escort-driver met us as we emerged from the restricted area of passport control into the breezy, exciting airport. Luke was a tall, handsome man of British descent, lean as a totem pole, fluent in Swahili, and with the knowledge of every African game habitat permanently mapped on his brain. He had escorted Juliet and her party on safari the year before and knew exactly how she liked things to go — easy and unrushed. I liked him at once for the boyish grin with which he welcomed Lou and me to the party, as if he found us to be everything the safari escort hopes for — young, beautiful and as shapely as a pair of Shirley MacLaines. I guessed his age to be in the early forties and I thought him quite brave to take on three agile, determined and long-lived ladies, any one of whom in other circumstances might have been his mother.

On the way out to Africa, Juliet had told us the few facts she knew about Luke's background which, on her previous safari, she had learned from him or his colleagues, mainly the latter, for Luke was not much given to talk about himself. He was born in South Africa, of European pioneer stock, his father a schoolmaster. In 1952, Luke had migrated alone to Kenya in response to calls from the British Civil Service, in which he served until Independence ended the British rule in 1963, some ten years after the Mau Mau revolt, during which he had been on active duty in Kenya's wild Northern Frontier territory. He was married and the father of two girls, whom he sent out to England for their education. He deplored trophy hunting and had never taken out a shooting safari. On my first meeting with him in the noisy, crowded airport he wore a dark business suit and white shirt with conservative cravat — a far cry from the bush-jacketed "White Hunter" I had expected. I soon learned, however, that he didn't have to dress the part: he was the safari escort in every fiber of his being, a tall, protective presence of quiet authority that had the native porters jumping to serve him, in response, apparently, to nothing more than his brotherly nods and charismatic smile.

The only missing attribute to complete the safari picture was the British Land Rover in which we would travel. Luke had brought his own car, a classy British compact, to drive us in style to our hotel, our luggage going in separate transport in the care of a hotel porter whom Luke obviously knew and trusted. The sudden freedom from baggage worries gave me a wild, carefree feeling. Safari travel, I thought, could be ruinously spoiling.

One of Nairobi's principal attractions, not mentioned as such in the guidebooks and seldom noticed at first by tourists because unexpected and invisible, is the quality of its high-altitude air. At fifty-five hundred feet above sea level, the city's air stimulates subtly, like a fine white wine properly chilled. Luke's driving seemed a bit wild until I realized he was keeping to the left, a holdover custom from the time of the British rule. Landscaped parks flashed by, riotous with Bougainvillaea in bloom — dark reds and royal purples massed together with salmon pinks, all trained to tree shapes taller than our car. "I thought Hawaii had the spectacular Bougainvillaea!" I cried, and Juliet, unable to turn around in the low bucket seat, shook her head vigorously and said that this was an unanticipated stroke of luck for us all, to have arrived at the peak time of the blooming of Kenya's Bougainvillaea, which surpassed any she had ever seen anywhere.

The Intercontinental Hotel suggested a Waldorf skyscraper, handsome and modern in design, but once inside its luxurious lobby you felt yourself to be already on safari. Tourists in bush jackets, slacks and shorts were all heavily hung with cameras and binoculars and I even spotted a topee as we struggled through the melee toward the registration desk, following Luke, who carried all our permits and reservations in a worn leather portfolio, the hallmark of his profession of safari escort. All that we had to do was sign our names on the room cards, which was just about all I was capable of doing in the high excitement of realizing I was actually *in* East Africa after having talked, read, and thought about it for nearly a year.

We were to have two days here — time to catch up with the thirteen hours time difference between Hawaii and East Africa, time to visit Nairobi's famous Coryndon Museum, Snake House and Animal Orphanage and (between these study trips) to pack and repack the single suitcase each must live out of for the next month. The safari routine began that first evening when we met Luke for cocktails in the bar. With our first drinks in Africa, Luke administered our first doses of quinine. He had laid out the tablets on cocktail napkins, two for each, and we had to swallow them in his presence.

"From now on out," he said with a grin, "Friday is quinine day." Every week he would dole out our two tablets and watch them go down our throats, since he

7

couldn't trust us to take them later in our rooms. "*I do the remembering for my clients . . . there's a lot of ma-laria out where we're going.*"

The two days in Nairobi flew by. We did our sight-seeing in the Land Rover, learning how to haul ourselves up into it without Luke's ready helping hand and how to get out of it, one foot first gracefully toeing for the ground some three feet below the floorboards, then slip-ping the body down to that secure stance. We practiced a few shots of local scenes with our cameras resting on the windowsills, over which sandbags were hung to give steadiness. All this was old hat for Juliet, but she en-joyed seeing Lou and me get our Land Rover legs.

Between museum and animal park visits, we roamed the Nairobi shops, eying the shelves to see if they held any essential items we had forgotten to bring from the States. The shelves held everything for a complete sa-fari outfit, from bush clothes and hats to soaps, shaving creams, liquid hair shampoos, and every brand of British pharmaceutical from throat lozenges to laxative salts. Cameras, guns and camping outfits were also on display, priced not too steeply in Kenya shillings, which were then about seven to the dollar, roughly seventeen cents.

On our last afternoon in the hotel, there was no com-munication between our connecting rooms. We were all busy packing and repacking the single suitcase each must live out of for the next month, the bigger suitcases with town clothes already in storage until our return. From now on, Juliet had told us, we were each strictly on our own. If we forgot anything, we'd have to do without. . . .

8

The ladies and the Land Rover

Thoughtful packing was never my forte. I worried over slacks, sweaters, dresses, sportshirts and underwear, wondering which to put in first and which to wind up with on top, the article I might want first. With envy I saw my roommate Lou, the orderly trained nurse, finding a place in her suitcase for everything and memorizing where she put each essential item, including her heart medicines. I couldn't find enough corners for my own medicines, vitamins, cosmetics and camera film (thirty rolls of color, one roll of thirty-six exposures per day, as I had calculated). Then there were the books. It tore my heart to have to leave behind old loves like Isak Dinesen's *Out of Africa* and Hemingway's *Green Hills of Africa* to make space for the bulky field guides describing the mammals and birds, and my ring-binder diary with a hundred clean white pages in it, plus an extra pack of one hundred, just in case.

I had a canvas camera bag that would ride at my feet on the floor of the car. I had rehearsed the packing of my camera equipment innumerable times — the Nikon F with a 50 mm lens, a wide-angle lens in its case, an 85 to 205 mm zoom lens eight inches long, which I had taught myself to hand-hold without any quavering, and a collapsible tripod in zippered leather case. Each time I packed that canvas bag, I found more space left over for essentials I might need during the ride — small things like a folding hat, a sweater, extra cigarettes, a pocket flashlight. . . .

While all this selecting and rejecting, weighing and cramming was going on, I made periodic rest trips to the balcony of our hotel window to look at Africa from seven floors up: the sweep of the Athi Plains beyond the city, the gentle rise of the Ngong Hills twelve miles to the southeast (Dinesen's hills, a suburb of Nairobi now named Karen) and the distant brown threads of roads puffed with dust where animals or vehicles were moving. Tomorrow we would be on those roads, heading out for the bush!

Between the immense vistas of earth and sky, scores of fork-tailed black kites rode the winds. Their high-pitched wavering calls of excitement made my own heartbeats seem shrilly audible.

II

THE THEATRICAL ARK

OLD AFRICA HANDS squabble amiably about where to take newcomers for their first night in the bush. Though our itinerary was already fixed, our hostess fretted aloud from the front seat of the Land Rover — should she have scheduled The Ark first or maybe better have saved it for a final treat?

The Ark was for me the smashing curtain raiser to all that was to follow. Newly opened that year in the Aberdares, not very far north of the famed Treetops (where Princess Elizabeth learned in 1952 that she had become Queen of England), The Ark was exactly what its name and architecture implied — a ship-shaped structure where men and beasts met in intimacy with only plate glass to separate them. The sea around The Ark was the green montane forest of the Aberdares, seventy-six hundred feet up slopes that often reached as high as thirteen thousand feet.

We left Nairobi around nine in the morning of September 5 for the three-hour drive north through Kikuyuland to the Aberdare Country Club, the transfer point for The Ark and a mere ninety-eight miles for our first day in the Land Rover. Since most of the road was smooth macadam, we were not yet aware of how vulnerable is a coccyx.

Mount Kenya appeared on our right. Clouds were blowing across its ice-coated twin peaks seventeen thousand feet up into the blue sky, and when they parted, Luke stopped the car, and our three cameras clicked. He never had to be asked to stop, he always knew exactly when and where. Sometimes he would stop and we'd cry "What?" and he would point to a pair of giraffes loping over the savanna, too far to photograph but fascinating to watch through binoculars. I fell in love with giraffes from that first view of them in the distant wild. In reality, I developed such a visible hankering for them that when my companions spotted some before I did, they never said to me, "Look . . . giraffe!" but, "Look . . . some *more* of your friends!"

Curiously, they seemed to approach and show themselves as if in response to my longing to see them again. Watching them, I always recalled how Isak Dinesen had described them in her African classic: "I had time after time watched the progression across the plain of the Giraffe, in their queer, inimitable, vegetative gracefulness, as if it were not a herd of animals but a family of rare, long-stemmed speckled gigantic flowers slowly advancing."

As we rolled along with hardly a jolt and only a subdued metallic rattle from the Land Rover, Luke instructed Lou and me on some of the do's and don'ts of safari etiquette. No one, he gently advised, must ever get out of the Land Rover when it was stopped until he had done so first and circled the ground and shrubbery

14

around to make sure it was clear of dangerous animals or snakes. No one must ever try to photograph a native on the sly. Permission must be given first by the native and his palm crossed with a shilling or two. And when we would be on "game runs" in the national parks, we must keep absolutely silent when the driver maneuvered us up to the animal to be photographed. The wildlife on the reserves was accustomed to the sight of cars coming close but would bolt at the sound of the human voice. And of course, within the game reserves, no one, not even the escort-driver, was allowed to get out of the car.

What a delightful switch that would be! I thought. We the human beings confined within the windowed cage of the Land Rover when in the domain of the free-roaming animals. We to be the zoo creatures the animals could stare at if they found us worth a single curious look.

In the late forenoon we left the plains highway and entered the wooded foothills of the Aberdares on a secondary road where violent bumps soon taught us what the handrails in front of our seats were for. Before long, the Aberdare Country Club, unexpectedly luxurious on the top of a hill, appeared before us. This was our transfer point for The Ark.

Its immense courtyard was crammed with Land Rovers, jeeps, zebra-striped minibuses and private limousines. Piles of luggage tagged for the tour parties blocked the footways to the club entrance. Outcoming guests from The Ark met us incoming ones head on, everyone a bit wild-eyed, as in an airport when too many planes come in at once. Luke made a way for us through the turmoil into the clubhouse lobby, where the small overnight bags of the ingoing were lined up on the floor, already tagged with Ark bunk numbers. He went to the desk to register us, saying to wait for him in the bar. "We've got a good half hour until lunch is served," he said. "Please order a beer for me — Tusker if they have it."

We were now in the channel for The Ark, every move to it or from it timed with military precision: lunch in the club at 12:45; cars for The Ark leaving exactly at 2:30; dinner, night and next day's breakfast in The Ark; lastly, transport back to the club at precisely 8 A.M. Since The Ark accommodated exactly fifty-seven persons per night, this was the only way its aspiring guests could be handled. (And only the *fore*sighted aspirants, I might add, who thought to reserve their Ark bunk many months in advance.) The order underlying the seeming chaos astonished me — such an unexpected efficiency to find in the middle of Africa! I watched uniformed native club attendants skillfully riding herd on the laggards while I sipped an excellent Bloody Mary.

Only the hotel jeeps and minibuses, each in the charge of an official guide, were permitted to drive the last eleven miles up a densely forested road to The Ark. It was a one-way road with an air of hushed privacy. Nobody spoke in the car I rode in, separated from my companions. The spell of silence that Africa laid on even the most garrulous travelers apparently affected everyone, not just my own party that seldom wasted breath on small talk. Everyone around me stared out the windows watching for birds or beasts and silently pointing when one was sighted. It was like a dream passage through an exotic jungle scene of Henri Rousseau, every

16

glossy leaf of the primeval forest painted clear and shining in the midday light.

We came to a halt in a small clearing. This was as far as cars could go. The last hundred yards up to the still unseen Ark were made, very appropriately, on foot. Single file we climbed a steep pathway to a rustic bridge that spanned a deep gorge. There we suddenly saw the great, swaybacked roof of The Ark, propped high on stone pillars at the edge of a glistening mountain swamp — the most dramatic apparition I had ever beheld anywhere. A long, railed gangplank led to it, across a drawbridge that bore a sign to warn that it would be hauled up at six-thirty every night and not lowered again until the same hour next morning. Once aboard, there you stayed.

Once aboard, you became a child again as you explored the neat, nautical compactness of this arklike viewing place for the wildlife of the montane forest. Though no children under eight years were accepted, you did not miss them. All fifty-seven adults aboard wore a childlike look of wonder as they sought their tiny bunk rooms down passageways that ran amidships from bow to stern.

Lou and I shared a double — two bunks with just room between for one at a time to pass. Hot-water bottles had been tucked between wool blankets and beneath wall night-lights was a buzzer that would, if you did not turn it off, waken you at any hour of the night if interesting animals appeared at the wateringhole.

The outside areas of The Ark were given over completely to glass-enclosed viewing lounges, open observation verandas, and on the ground floor, an eye-level pillbox for photographers seeking special close-up views. In the midships bar, where a great log fire burned all night, there was a daily logbook recording not winds and tides but the number of animals seen at the wateringhole. This was the logged record for the day preceding our arrival:

ANIMAL	BY DAYLIGHT	AFTER DARK
Bushbuck	1	1
Buffalo	1+2+1	6+2+7
Warthog	1	
Elephant	30+2	
Rhino	1	2+1
Giant Forest Hog		9+7+3+8
Genet		2
African Hare		4

The stage for these appearances was a great, saucer-shaped swamp with trails leading down to it from the forested slopes that ringed it. The front of this natural amphitheater faced the windows of the observation lounges. White salt was scattered around the mudholes there, just behind the footlights, so to speak, a few yards from where the audience sat perched above, behind glass, on comfortable leather lounges with armrests to hold cameras, binoculars and liquid refreshments while waiting for the actors to come on.

The elephants came first, not long before sunset. One by one they stepped out from the distant wings of a bamboo forest and came slowly down the trails single file along both sides of the swamp. The sun sent its glow to their wrinkled hides and flapping ears. Their bulk grew bigger with each step down toward the salt until,

The Ark in the Aberdare Mountains

The Ark — the elephants came first

in the magnification of my lens, I could only get portions of them at a time — a head or a rump or a tangle of trunks and huge feet tipped with stumpy toenails that pawed at the salt, which trunk tips snuffled up greedily or, more endearingly, laid bare for the baby elephants that now and again walked into my lens from under the shadow of their parents' ponderous flanks. I hardly breathed as I focused and clicked, opening my lens wider and wider as the daylight failed, until I was down to the last wide-open stop at 3.8. When it was over, after perhaps an hour or more of the most tense excitement, I was drained of all feeling, as one emerging from a passionate love affair.

Slowly the elephants ambled along the swamp edges, returning to the wooded backdrop, into which they disappeared, melting into it without seeming to disturb a single branch. It was magic to watch them through binoculars. A few old matriarchs remained front-stage, on their knees now, digging deeper into the red, salty mud with red-stained tusks splintered at their tips from years of such violent excavating. And before the last elephant had left the stage, the first Cape buffalo appeared on it, then a second and a third.

Massive, black and taurine, their heavy horns grown together in a bony boss on their foreheads, they looked to be exactly as reputed — the meanest and most dangerous beast of the bush. They gazed at the mess the elephants had made of the wateringhole, but then found places to drink from, where fresh water was already seeping into the deep footprints.

Though I had had it, photographically speaking, I tried one last shot of a ton-weight bull buffalo standing out front just a few yards away, facing us with horns lowered and mean eyes threatening, as he caught the smell of man in the pillbox below, where cameras were waiting in open slots of the thick stone wall. The buffalo stood, a motionless statue of dark menace, debating whether to charge, then he turned away casually and cropped some marsh grass, the picture of bovine innocence.

There was an expired "*Aah!*" from the viewers around me and I became aware of people again and dropped back into the present from some distant age, the Pleistocene, perhaps, when mastodon and mammoth owned the earth. I looked around for my companions and saw Luke handing them each a highball. He brought me a third tinkling glass.

"I'm not back yet," I told him.

"Drink up, sister," he said, smiling. "This will help bring you back."

Darkness was falling over the swamp when we heard the call to dinner. Not a black darkness, but a peculiar silvery radiance from a fifteen-thousand-watt "artificial moon" that had been lit outside the observation lounge. Its diffused beam covered the entire watery stage.

Reluctantly, we tore ourselves away. The dining hall was down in the hull of The Ark — a long, windowless room offering no distraction beyond what lay on the lamplit tables. With what care it had all been thought out! First, unique visual impressions to feed the soul, and now this feast for the body. The menu (one for each guest as a souvenir) was hand-printed on a strip of bark cloth dyed reddish like the mud of the wateringhole, decorated with black-stamped animals of primitive de-

21

sign copied from East African cave drawings. The menu offered:

Seafood Cocktail
Oxtail Soup au Sherry
Filet Mignon au Parma
Pommes Frites
A variety of Salads
Iced Fresh Fruit Salad
Kenya Cheese & Biscuits
Coffee

The look on my face as I read it doubtless prompted Juliet to remark: "The Kenya cheeses are famous and excellent" — to remind me that I really *was* in the middle of Africa. *Kenya,* she said very clearly, enabling me to locate myself geographically. But that was only a lost sense of place induced by a menu of gourmet foods (how the devil did they ever get them *in* to this isolated water-hole?) presented with artful sophistication.

Out once again on the observation deck, the sense of both place *and* time vanished completely. Impenetrable darkness now surrounded the disk of artificial moon-glow. Nearby plaques of swamp water glowed faintly, suggesting primeval ooze preparing the first living thing that would perhaps in a million years crawl forth from it. From some placeless point out of time, one seemed to be looking down on the creation of the world before God said, "Let there be light."

Then the first of the night prowlers began coming into the spotlight. . . .

Sleep seemed all but impossible that night, not just

for novices like Lou and myself, but for most of The Ark's inhabitants. Like somnambulists they moved through the dim corridors that led out to the observation lounges, back and forth from bunks to viewing place — a ghostly parade of sweater-wrapped forms soundless in bed slippers.

When not on the prowl for "one last look," I lay awake in my bunk listening to the medley of quacks, honks and thin, reedy whistles from the waterfowl in the swamp beyond that penetrated the small, steel-screened porthole above my head. Earlier, an ornithologist aboard had identified them as Egyptian geese, African black duck and some crowned cranes. They all took longer to settle down than I did.

III

NORTH TO SAMBURU
GAME RESERVE

Next morning, looking a bit like revelers hung-over after a night on the town, we were driven back to the Aberdare Country Club and our own vehicle. The Land Rover now had a homelike air, with all the duffle we had not taken up to The Ark stacked in the rear compartment, our bush hats and Kenya road maps left on our seats and, underfoot at my place, some ground-out cigarette butts that I had forgotten about in the excitement of my arrival the day before.

"This will be your first real workout," Luke said. "It's about a hundred and twenty miles up to Samburu." The highway stretched far before us, restfully vacant of man or beast, nothing to look at except Mount Kenya on our right, its crest invisible under morning clouds. I told my companions that I doubted I'd be able to take in a single new impression that day. "Not even a giraffe," I added to underscore my feeling of surfeit.

But then I saw the weaverbird nests in the open acacia woodlands we were driving through. They were golden globes made of straw, six or eight inches in diameter, hanging in clusters from spiny branch tips.

They were woven upside down, Luke explained, to protect eggs and chicks from all predators except a certain
tree snake. He stopped under a roadside tree so we
could look up at the tiny nest entrances — dark holes
at the base of the globes, like small, round mouths agape.
The wonder of this avian creation overwhelmed me. I
reached for my camera and went to work. Africa, I was
discovering, gave you no interims.

The highway all the way up to Nanyuki and beyond
might have been named Weaverbird Road instead of
#816. The globular nest colonies swung overhead by
the countless hundreds, not from every acacia tree but
only from certain ones which, to the human eye, looked
exactly like the untenanted trees. Winds apparently had
nothing to do with site selection. The winds blew from
every direction across the highlands and when they
attacked the nests, the straw balls danced in circles from
their slender moorings. But nothing ever was shaken
out of their bottom-side entrances, no eggs, no chicks,
not even a feather . . . a mystery compounded the
enigma of those nests.

What individual weaver first thought of building its
nest upside down for predator protection, somewhere
back in the dawn of bird-time? What had happened in
that one small bird brain that had led to this revolution-
ary new style in nest building? It had to have happened
with just one, at first. One weaver with one clever idea
fancifully worked out in bird-thought before proving
the practical application. . . .

The field guide to the birds of East Africa (always
kept at hand on the car seat) stated that the weaver and
allies comprised one of the largest bird families in Africa.

28

Maybe *that* was an answer in a way. Those predator-proof nests had enabled the family to flourish and multiply, although its individual members were as small and defenseless as the true finch, which they strongly resembled.

Besides giving no pause in new impressions, Africa also does strange things to your customary thought patterns, pushing you into deep pondering on the Beginnings of Things. The antic customs of birds and beasts that had originated in a dawn age acted themselves out before your eyes.

The crossing of the equator just before Nanyuki brought lighter thoughts. Here in the middle of a great, empty plain stood a signboard with the map of Africa painted brown on orange and a band of crimson reading EQUATOR bisecting the continent just under the bulge. "This sign is on the Equator" it stated, rather redundantly, adding "Altitude 6,289 ft." Beyond the sign were rows of sheds filled with handicrafts of the local tribes. Several long tables were solidly peopled by foot-high carvings of warriors with fur headdresses, spears and skin shields; herds of animals were grouped by species — gazelles, rhinos, giraffes, warthogs and others. Our first native market gave the feeling of meeting the whole of Africa in miniature. It was also our first opportunity to photograph freely. The wood-carvers behind the long tables sat on the ground amid piles of shavings and hardwood logs, their intent black faces bent over the pieces of wood they studied endlessly between cuttings.

After watching one carver concentrating on a slender log he was slowly shaping for a giraffe, I was not able

Nanyuki — Lou bargains in Swahili with a wood-carver

quite to follow Luke's advice not to pay the first price asked, but to bargain so as not to lose face. Why worry about losing face when you had already lost your heart? Native markets, I should explain, have always been my undoing; I become idiotic, speechless with delight.

It took me more than a half hour to decide which two, among some fifty carved giraffes, would make the most beguiling pair to own. No two were exactly alike, though all were similarly posed, with the small, horned head looking off to the right or left and the ears turned forward attentively. The mark of the carver's individuality was on each one, something subtle his knife had cut just a shade differently. And the wood stains they used to paint the markings on the giraffe hides were applied with differing densities of color! Only the urgent horn of the Land Rover brought me to my senses and forced me to choose, pay up, and pick up.

Luke let us out of the car again in Nanyuki, where he had to stop for gas, beer and lemon soda. He gave us a half hour furlough to roam the single main street lined with shabby shops, mainly of the trading-store type. I hunted for a bag to hold my fragile giraffe carvings, while Juliet and Lou went off on a more practical search — for Life Savers or hard-candy fruit lozenges to keep our salivary glands working when we reached the dusty roads of the desert that lay ahead. They found what they wanted but the only desirable bag I discovered, in an Indian shop, was priced at a prohibitive ten dollars. "And absolutely *no* bargaining," said the sharp-eyed Hindu in charge. She wouldn't even take the bag down from an overhead hook for me to examine. I walked out of that Hindu bazaar wondering where and how I had ever acquired the *idée fixe* that all Indians were charming and gracious. In contrast with the laughing hospitality of the black wood-carvers down the road, the Hindu shopkeeper seemed to be an alien presence in this sunny-natured land.

After Nanyuki we began climbing a mountain escarpment that circled the base of Mount Kenya. In a forest glade by the banks of a mountain stream jumping with trout, we had our first picnic lunch. Everything was still a "first" to me, an exciting adventure, and I never tired of what became the routine of that lunch stop, which set the pattern for all our future picnics.

First, Luke descended to explore the bushes, tall grasses and footpaths round about. When he decided where to set up the three folding camp chairs, he returned to the car to get them and to open the doors for us. This was the signal for us to descend. He then told us which bushes we could go behind, if need be, and turned his back to us while setting out the chairs. When we came back from the authorized bushes, each chair had a box lunch made up by the hotel of the night before, and the big red cooler box, filled with bottles bobbing in melting ice cubes, stood in front of the chairs, four plastic cups on its top. He poured for each of us the beverage chosen, then sat on his heels in front of us with his own box open on the ground. Usually we ate in silence, staring around at the strange trees, sometimes fruited, sometimes flowered, through which elegant, small birds flashed their streaks of crimson, gold and green, too swift for our eyes to identify, but not for Luke's. He named them for us in a low voice — spotted creeper, golden-winged sunbird, lilac-breasted roller,

carmine bee-eater . . . names that made music for our
fête champêtre.

In the course of our thirty days on safari, we pic-
nicked in the most unlikely places — on deserts under
thorn trees, on crater rims shrouded in mists, on stony
escarpments overlooking countless miles of Rift Valley
plains shimmering with white soda lakes. Every lunch
spot was a new adventure, a new sensation, a new bit
of Africa taken into oneself during an hour of sitting in
one place, gazing about, hypnotized and almost para-
lyzed by these singular marvels.

After this first lunch by the trout stream, we continued the climb up the escarpment through the forests
of cedar. At an altitude of some seventy-five hundred
feet, the road made a turn and we began to descend.
Here Luke pulled off to the side so that we could look
our fill. Now the plains of Samburu Isiolo Game Reserve
lay far below us, a vast, dun-colored wilderness that
took our breath away. Beyond to the north, behind shift-
ing mirages that resembled a silk screen blowing in the
winds, stretched the deserts of the Northern Frontier
Province. I could understand our escort's joking comment that from here one could see all the way up to
Ethiopia, if one looked sharp. This was another "first"
— the feeling of the wideness of space one received from
any high point, as if Africa went on and on forever.

We arrived at Samburu Game Lodge in the late afternoon after jolting over miles of unpaved roads that
wound like game tracks through thornbush close enough
to scrape the sides of the Land Rover. No signboards
directed where the lodge might be found in this dusty
drought of desolation, but Luke knew the way. It burst

suddenly into view like an oasis of green. Here were the banks of the Uaso Nyiro River and the game lodge — an enchanting spread of rustic cottages with palm-thatched roofs, a dining hall open on the river side and a great, circular bar built on pillars that lifted it out over the riverbank directly above the brown stream, where crocodiles were floating like discarded logs.

We began to explore at once, while Luke registered us and told the porters where to put our baggage. There were three cottages reserved for us — a double for Lou and me — and everything was unloaded, as we were to stay here for three nights.

Our three cottages stood side by side along a flag-stone pathway shaded by acacias and palm trees filled with birds. Bands of black-faced vervet monkeys watched us moving in, studying us intently as if deciding which of us could be counted on to bring them bread after meals, despite the signs all about begging PLEASE DO NOT FEED THE MONKEYS. They at once selected Lou as their pushover, as if they knew that she had once owned a monkey in the Congo and remembered well the delights of that simian companionship.

After showers and a change of clothes, we met in the bar, where our escort had captured a riverside table. The proper gentleman, he stood up as we entered and welcomed us with his boyish smile, touched slightly with surprise because we all looked so combed and clean, nothing like the dusty, windblown witches he had helped step out of the Land Rover scarcely an hour before.

Sipping our highballs, we stared across the river at the opposite bank, a circus stage occupied by a troop of

34

baboons come down to eat a bright green weed that grew at the water's edge. Behind them, half hidden in leafy bush, were some elephants. In the crotch of a dead tree standing mid-stage, an enormous hunk of dark meat was hung. This was to lure a leopard, Luke said, that frequented these parts and with luck we might see him that night.

Staring, staring . . . no matter where you were in the African wilderness, you stared about in wonder until your eyeballs felt unsteady in their sockets from over-work. Here, in the middle of the Samburu Isiolo Game Reserve, separated from its wildlife only by three hundred yards of river fronting the lodge and by a stone wall and guarded gate on the land side behind it, there was almost too much to take in. Words such as *plethora*, *exorbitance*, *surfeit* and *glut* ran through my mind as I stared, from our dress-circle box, at the royal opera going on across the river.

A female baboon in season was presenting her back-side callosities to an uninterested male and screaming at him, outraged that he was refusing the offer. Again and again she tried in vain; even tugged at his mane with her long gray hands. Her screams rose to a pitch of hysteria that stopped only when the male spun about and gave her a clop that sent her spinning down the slope.

Long after we were tucked under our mosquito nets that night, we heard her again and again, screaming in mounting furies of frustration. Had we not seen earlier the enlightening scenario of the she-baboon spurned, we might have thought a human being was being torn to pieces by a wild beast.

Elephants along the Uaso Nyiro River at Samburu

A pair of fifteen-foot-tall giraffes

Next morning we breakfasted early and were off before eight-thirty for our first game run — a titillating phrase in all safari advertisements that suggests a chase of some kind. Actually, the game run is no run at all, but a slow-going search in a Land Rover through open bush country, looking for the wildlife it conceals and stopping soundlessly when something is sighted. Then comes a creeping advance on quiet tires up to camera range and the maneuvers to get downwind if possible, and of course the right light for shooting.

For the game run, all windows and the two top hatches of the Land Rover are open, all cameras are held at the ready, propped steadily on sandbags hung over the windowsills and hatch openings — a most effective device in lieu of a tripod for long shots at moving subjects. A sort of genius-driving of a safari escort is required for this unique game run over roadless savannas full of potholes and dry creek beds, and of the passengers is demanded total silence, patience and, when someone is shooting, immobility.

The first animal we put up on our first game run was a fifteen-foot-tall reticulated giraffe. He was just outside the lodge gates, cropping the leafy top of a thorn tree. For about ten seconds after Luke stopped the car, he stood absolutely motionless, looking down on the strange metal contraption with queer white faces staring from the windows and roof hatches. Then soundlessly, without fuss or alarm, he stepped around to the other side of the thorn tree and vanished utterly. The dark geometrical patches of his handsome hide, marked off by narrow white stripes, merged perfectly into the sunshadow pattern of a tree in morning light. I got one shot of him before he became invisible, not because I was quicker on the release than my companions, but only because he had appeared on my side of the Land Rover.

Our remaining sightings that first morning were all beasts and birds new to Lou and me — gerenuk, dik-dik, warthog, Grévy's zebra, Grant's gazelle, impala, mongoose (on an anthill), a flock of vulturine guinea fowl (with brilliant cobalt-blue neck plumage), a white-bellied go-away bird (whose call really did say "Go awayaaa") and a black-feathered drongo with red eyes and forked tail.

When we turned back toward the lodge for lunch, I had run out of film, a blessing in disguise. I could at last just sit back and *look* at the extraordinary creatures I had already made mine in the safety of the camera. This was my deserved reward for a hard, sweaty morning behind a zoom lens. The elusive dik-dik, for example, looked smaller and much more fragile with the naked eye — a true antelope in miniature, barely fourteen inches high, with tiny, spiked horns and minute tufted tail.

Routinely, we were supposed to take a siesta after lunch, to recoup our energies for the afternoon game run. Lou and I found this difficult to do at first, though the big animals themselves set the sensible siesta pattern of lying up for rest from the midday sun. We spent our siesta time on the open front porch of our cabin, feeding the vervet monkeys (which apparently never slept), tossing crusts of bread with fast throws to keep them out of our laps and at safe arm's distance, since their bites, Lou warned, could give a nasty wound. When the bread was gone, they sat on our low stone wall, long gray tails hanging down to the floor, watching with liquid black eyes every move of our hands. The white band across their foreheads, blending at the sides with white whiskers, framed their triangular black faces, giving them a cowled, monastic look, secretive, sad and wise.

I read aloud about them from the mammal field guide while they gave us the pleasure of their company. Vervets weighed about ten pounds, were eighteen to twenty-six inches in length (without tail), lived in bands of from six to twenty, ate a mainly vegetarian diet supplemented by grubs, caterpillars and occasional eggs of ground-nesting birds, drank rarely, and were preyed upon by the big cats like leopard, serval and caracal.

"This is the way to learn the animals," I said, "taking them one by one — a study in depth." The field guide index listed over six hundred different African mammals, of which fewer than a dozen — lions, leopards, elephants, etc. — had been seen by us before, and then only in zoos or circuses. "One by one," I repeated; then we both laughed at the idiocy of the remark. The animals seldom came one by one out in the game reserve, nor would they sit peacefully before you as subjects for a "study in depth." There was only one creature that would do this, but we had not yet met the lion in the wild.

On our afternoon game run, Luke had driven us far out into the reserve and was running along the stony bank of a narrow stream, looking for a safe crossover point. We had seen no wildlife for quite a time and were unprepared for the sudden halt at an opening in the thornbush and Luke's whispered "Lion!"

At first I could see nothing because I was looking too far. Then I saw where Juliet was aiming her camera. The lions lay below us in a patch of shade on the opposite bank, no more than twenty-five feet from the Land Rover. "Honeymooners," Luke whispered as I stared down at them in a trance of unbelief.

They were a magnificent pair stretched out, one above the other, on the steep slope, the maned male on the high side and the lioness below him. The tip of her tail hung down into the stream and one of her front paws lay possessively over his hind legs, folded together beside her face. Both beasts appeared to be sound asleep in their honeymoon bower, and it was indeed a sight to behold.

When at last I could lift my camera, I discovered that their combined length in tandem stretched out to more than fifteen feet, and that I was too close to get both their bodies within my big lens. I had to snap them separately, beginning with the lioness and her tail that trailed in the river. When I raised my lens to the male, he suddenly lifted his maned head, looked down to his sleeping bride, then across at us with indifferent eyes. Soon he dropped his head back into the grass and closed his eyes.

We watched them for almost an hour, hoping they might bestir themselves into giving us some new poses. But they remained motionless, untroubled by the small sounds of our whispers and clicking cameras. Before long, the accumulating heat in the Land Rover impelled us to move on. "We'll find a breeze and maybe more lions," Luke said to console the two in his party who had seen their first lions living free and would willingly have stayed on and on for the sake of watching them.

The impact of my lion sighting lasted the rest of the afternoon and its force increased the more I thought about it. Lions encountered in the wild were totally different from any preconceived ideas about them. Nothing I had read or seen on television nature shows prepared me for their royal disdain of the Peeping Tom tourists, for their huge laziness and their frank honeymoon affection while resting between matings. All the ancient lore about lions that stressed only their savagery now had to be revised.

I realized, of course, that if one of us had fallen out of the car when viewing them, it would have been a different story. Then we might have seen the clawed killer that Hercules wrestled to its death in the Valley of Nemea or the roaring fury early Christians had known in Nero's Roman arena. Yet, even in their seeming gentleness, as they slept by that stream, I had felt their mysterious power.

Before the end of our safari, I was to see uncounted lions in all sorts of places, moods and postures — panting aloft in trees, growling bloody-muzzled over a kill, stalking a meal across dry grass plains. Yet never would I be able quite to analyze the essential mystery of the lion's presence — the power to capture and hold the human gaze indefinitely, even when doing nothing but sleeping. It was an inexplicable kind of animal magnetism that affected everyone — natives, seasoned escort-drivers, even the bored and trivial-minded tourists. I

41

never heard anyone say in Africa, "When you've seen one lion, you've seen 'em all."

No two lions were alike, no two lion personalities were alike. But *all* possessed that singular power over their human observers, the magnet that could hold their fixed attention hour after hour. Even the cubs had it.

IV

NATIVE DANCES

For our second day at samburu, Luke had kept a surprise in store for us. He had learned from another safari escort that arrangements had been made for his party to see the dancing of the Samburu warriors in a village out on the edge of the reserve. Luke had contributed a share of the cost asked by the tribesmen to dance for "outsiders," and thus had arranged for us to join the other party, in our Land Rover.

After lunch, we set out from the lodge with our cameras loaded — Juliet for movies, Lou for black and white, and I for color stills. No matter what happened to the sunlight, one of us was bound to come back with some gem captured on film.

Beyond the green belt of the Uaso Nyiro River, we came upon a dry, stony plain that even the wildlife avoided, above all at midday. The empty, dead land stretched out to the horizon, like an Arizona desert minus the cactus. I asked Luke what kind of people could live in such desolation and he pointed to the answer — a nomad boy with a small herd of goats moving in the direction we were going, toward the village.

The boy carried a slender staff twice as tall as he was and his thin, black body, naked save for a strip of cotton over the shoulders, looked from afar like a dried leaf hanging from the end of a branch. But as we were passing him, he came alive and with great animation answered Luke's "*jambo*" with a return greeting spoken twice in a boyish voice shrill with vitality, "*Jambo . . . jambo sana!*" and a wave of his long herding staff. Before our cloud of red dust enveloped him, I saw with relief that he wore tattered sandals to protect his feet from the hot, sharp desert rocks.

The Samburu village beyond lay on the top of a hill overlooking still another great expanse of stony desert with a horizon of blue mountains. At first sight, it seemed to consist only of four great rondavels, like watchtowers with peaked, thatched roofs, high above the road to the left. To the right was the game warden's rondavel, where Luke stopped to check us out of the Samburu Game Reserve and get our papers stamped for free entry back into it on that same day.

We got out for a closer look at the warden's gatehouse — a handsome example of rondavel architecture with hand-hewn beams supporting the peaked roof. Above the bench around the rear wall, lettered in gold on a plaque of native wood, hung a sign:

In Memory of
ELSA
Who Helped Safeguard
This Game Reserve

So this was where some of the *Born Free* book royalties had gone, right back into the game lands where the famous lioness lived with Joy Adamson like a member of the family. It was a touching tribute from the young black Republic of Kenya to a white writer whose books had brought the wonder of their lions in words and motion picture to the entire world.

"That's immortality for a writer," I said as Lou photographed the sign. "To be remembered like that, out here in the middle of nowhere. . . ."

Natives began gathering about the gatehouse to look us over with side-glances that feigned lack of interest. These were the Samburu, a tribe related to the Masai people. The men are tall and narrow-hipped, their hair braided into tiny pigtails and smeared with orange ocher-grease; the women shave their heads and crown them with bead coronets. Their bodies are hung with coiled wire studded with crude gemstones that abound in this region. I kept the lens cap on my camera as I wandered among them, for these were not the dancers we had paid to see and to photograph.

Luke finished his paperwork in the gatehouse and drove us down the hill to the dancing place — a quarter-acre stone pasture on a slope covered with a fine sand that bore a thousand footprints of men and cattle and was surrounded by the huts of the real village, low, round-topped bomas made of mud and wattle that so merged with the thorny landscape as to be unnoticed until you were close upon them.

At the top of the slope were the parked Land Rovers of the other safari party, six in all, each with a guardian to keep off the naked children looking for loot. The

Practice jumping of the Samburu warrior dancers

Samburu warrior dancers

dancing place was crowded with greased warriors huddled in a group at the base of the slope; the shaven-headed women were gathered beside them and all were chanting and laughing together, as if alone in the world.

The dance master, six feet tall, wore a rumpled old army jacket and a white-sheet sari. He came up to Luke and apologized for the dancers being late in starting. It seemed they didn't have enough ocher-mud to decorate themselves properly and the runners sent miles away to fetch it had not yet returned.

"These dance things go on all day," Luke said. "We can only hope they'll get going before sundown."

I didn't care if they never got going. It was enough simply to be there with fifty orange, greasy warriors before me, practicing their leaps within camera range, their women shouting praise and clapping their hands in unison. There was no music of any kind, nor any drums to keep the beat, just the sounds the women made. The men's orange cotton loincloths and orange hair and faces made a shout of color against the blue sky when they leaped, the sharp metal points of their long spears shining above. When I zoomed in for close-ups of the women's faces, I saw the unbelievable beauty of the young girls. Their proud, rounded profiles above slender, long necks were like dark dahlias in the sun. Only occasionally could one catch a profile and seldom a full face, for they turned their backs to us with studied disdain.

The bright groups of dancers were surrounded by the ordinary inhabitants of the village — women in brown cotton garments, their babies slung in looped cloths hanging from their backs. Swarms of boys naked except

"The unbelievable beauty of the young girls"

for a string of beads mixed with old crones as thin as
thornbush branches and as dangerous to brush against.
One of them caught me aiming my camera at her and
came after me with a stone in her hand, her thin black
arms flailing out like a baseball pitcher winding up for a
home-run hurl. I snapped her as I backed away and
caught her face of rage yelling at me in Swahili.

The nondancing natives interested me much more
than the elaborately decorated dancers, especially the old
black grannies with their withered, wise faces and
straight backs. But I could not persuade one of them to
stand willingly for her picture, nor could I sneak a shot
even if one had her back turned to me. It was as if they
all possessed peculiar antennae that told them when a
camera was aimed at them. Instantly they would turn

about and make a threatening gesture or walk rapidly out of range. They appeared to be the guardians of the privacy of their people, fiercely intent on keeping every aspect of their lives hidden from the pale photographers and to hell with any deal their chief had made with intruders!

The dancing started down the slope with a chorus of grunting shouts from the warriors as they began their leaps straight up in the air while the women dancers clapped the beat with arms and bracelets held high above their shaved heads. It seemed no more than the practice leaping I had photographed earlier in a better light, except that now the leaping was continuous and even higher as the warriors competed in pairs, trios and quartets.

Although monotonous and repetitive to our Western eyes, this continuous leaping fascinated the villagers. They stood transfixed between the dancers and the visitors, understanding something about those repetitive leaps that was unknown to us. Even the crones stopped their protective prowling among us to watch their sons and grandsons soar straight up from the ground, arms held tightly at their sides, seeming to have mysterious coils of steel as muscles in their bare feet, which shot them aloft over and over.

"Will they do something else?" I asked Luke.

"No," he said, "this is it. They'll go on jumping till the sun goes down, and afterward into the dark."

We drove home to the lodge by a new route that led us back to the Uaso Nyiro River just after sunset, when the wildlife was coming down to drink. Along the riverbank on the Samburu side we put up our first cheetah.

She was sitting erect on her haunches, a spotted statue of feline elegance, her black ringed tail curled about her. Her gaze was fixed on the riverside shrubbery with a concentration almost electrical in its tension, and so she remained motionless in profile to us. Then she turned her small, haughty head and let us see the characteristic teardrop markings of her golden-spotted face before she stalked off without haste into the palmettos.

Nobody in the car spoke a word for the rest of the way back to the lodge.

V

BUFFALO SPRINGS
AND
MORE HONEYMOONERS

GAME RUNS TWICE A DAY never became merely routine. You never knew beforehand what new animal might be put up. Each run was a gamble in adventure, suspense and excitement, especially in the Samburu Isiolo reserves, which had ten miles of permanent river water running between them, enough to support (according to the field guide) nearly one hundred different species of mammals, from bats to hippos and all assorted sizes in between.

On our last day at Samburu we took off early in the morning for a run up to Buffalo Springs on the Isiolo side of the river. Luke had added a game warden to our party for this run, a slender young black in a khaki uniform and green kepi. He carried a gun and rode in the rear compartment of the Land Rover with his roof hatch open. From there he watched the road ahead and called out in Swahili to Luke words that meant right or left when we came to a fork. At intervals I stood up on my seat to look out the open hatch, and then both sentinels, the warden and I, became two busts, one black, one white, watching intently the open bush. Occasionally I passed him a cigarette, which I knew he was hoping for.

We were far out in the open bush country when Luke raised his hand and pointed to a dusty grove of scattered thorn trees. "Oryx," he whispered . . . "*beisa*, I think." Only his keen sight could have detected the gray antelope bodies at this distance, but as he cautiously advanced the Land Rover, we all saw them at once when they turned their black and white masked faces toward us, with those long, parallel horns growing straight up from narrow foreheads, sharp as rapiers. When the motor was turned off, they ignored us, continued grazing and trotting back and forth, which gave us splendid views of their four-hundred-pound bodies in motion. Now and again a pair of bulls lowered their yard-long horns and went at each other head-on as if for a fencing match. The crashing clacks of horn against horn we clearly heard at thirty yards' distance. Luke told us in a low voice that in rutting season these matches were really savage and what we were seeing now was only practice play.

We'd had luck, he said, to run into this herd of more than forty oryx living their sportive lives as if in their natural privacy. We lingered for an hour, watching them through the camera lenses, clicking every time we caught full-on a white face with its handsome, sharp markings in black, masks for a ball of prancing and dancing. After we drove away, that's what I felt I had been watching — a masked ball of the great antelopes, *Oryx beisa*, in a thornbush glade.

The stony track led us northward into another desert wilderness of thorn scrub. The sun beat down hotly through the open hatch and Lou and I put on our bush

The masked faces of the *beisa* oryx

Her round yellow eyes were like polished topaz

hats to please Luke, who was worried about us and said something about mad dogs and Englishmen. He and Juliet on the front seat were protected by the fixed roof of the car.

"Tomorrow, when we cross this desert to get to Marsabit, the hatches will be down," he said.

Meanwhile, we three could take it straight in the face, Lou, the game warden and me. We stood on our seats, our elbows at rest on the hatch coamings, and received directly the great vistas of northern Kenya. This, to me, was the most wonderful way to discover Africa, dusty mile upon dusty mile, as if leaning on the sill of a moving window that had no limiting frame.

Until the game warden spoke the name, we were unaware that we had come into Buffalo Springs. We saw some scattered sheds with pieces of broken machinery and a few seemingly abandoned thatched huts and as Luke wrestled the Land Rover down a stony incline, the game warden suddenly said: "*Simba!*"

Simba, the lion! Luke swung us off the track and circled the base of an enormous acacia tree, driving over roots as thick as thighs coming from the stony ground. Between two great roots splayed out like a V from the main trunk, he stopped the car so we could look down on a pair of honeymooning lions dozing in the shade not twenty feet from us on opposite sides of the tree's main trunk. The male was curled up like a giant yellow cat, his head resting on four paws, asleep. The lioness on the opposite side sat relaxed, head up and eyes open, keeping watch for her mate.

We moved in slow motion as we reached to the floor for our cameras and the sandbag "tripods" to steady them for slow shots in the shade. Luke kept the motor running for safety, just in case . . . a low purr of the engine in neutral. I thanked God that I had kept the zoom lens mounted on my camera. In its magnification, every whisker of both lions stood out clearly, and when the lioness turned her head and looked straight down my long-barreled lens, her round yellow eyes were like polished topaz.

The lioness now communicated in some mysterious, soundless way with her mate. He opened his eyes and blinked. The low whirring of Juliet's movie camera and the hum of the idling motor were the only sounds in the otherwise absolute stillness. We kept our panting breaths silent and shot as fast as we could. Presently the lioness stood up, looked behind the tree at her mate. Her signal seemed to say, "Let's get going!" When she started to walk away toward the village, he slowly stood up, yawned, and followed her. Side by side the tawny bodies moved into the bushes and instantly disappeared. At last I sat down and breathed freely. The film counter registered fifteen exposures in those long minutes of purest wonder.

Luke drove us to the flatlands that surrounded the oasis of Buffalo Springs. At first sight it looked like a Hollywood set complete with shouting actors bathing nude in a pool. How astonishing to come upon such a scene in this wilderness!

Here were two couples of young Europeans, all with long hair and tanned bodies, very free and beautiful, without the restraining influence of an official guide, required only when inside a game reserve. Their battered jeep was placed under an acacia tree and their shorts, halters, and miniskirts were scattered on the ground between jeep and pool. Luke parked beside the jeep and went to the pool to warn the noisy bathers that there were lions nearby.

Their reaction was to cheer him, so eager were they to see those lions. They ran, dripping, to their jeep and from their cries we never knew what language they spoke or from which direction they had come, for within a few minutes of Luke's announcement, they were off up the slope to look for the lions, their jeep engine belching smoke and clattering like a machine gun.

"I'm glad we got to those lions first," Juliet said.

"They'll never get within miles of them," said Luke and passed us our beers and lemon sodas. Now we could sit in sweet silence and gaze at the beauties of Buffalo Springs. The sky reflected in the pools was cobalt blue, framed in green rushes. In this pale wilderness, the pools contained the only colors they knew — brilliant blue and green. A single palm tree grew from the source of the springs, a slender trunk that separated into three branches, each carrying a crest of fronds.

"Take a good look at that palm," Juliet said. "It's the only branching species of palm on earth, native of Africa, called the doum. First time I saw them was in Nubia. You can imagine how excited I was, coming from Hawaii with the fixed idea that all palms grew straight up, branchless. . . . It was my botanical first."

Though an experienced botanist, she called herself an amateur, yet her great store of botanical knowledge, coupled with a passion for every growing thing, weed or tree, enriched every scene we shared with her. And now

Luke and the park guide resting under a thorn tree

by saying "the only branching species of palm on earth," she had added a touch of magic to the singularity of Buffalo Springs.

"Do you think it's a survival form from an earlier age, Juliet?"

"No. . . ." She looked up at the doum palm as if to question it, then added slowly, "I think . . . maybe . . . it's a *development*."

Luke and the game warden were already waiting in the Land Rover. Juliet, Lou and I collected our individual cameras and binoculars, gave each other the usual spot-check inspection to be sure that nothing had been left behind (a protection routine because we were all somewhat absentminded), and climbed aboard.

While passing through the village, we kept a sharp watch on the shrubbery in the foolish hope for one more glimpse of our honeymooning lions, and out once again on the stony plain we looked in vain for a telltale puff of dust that would mean the rollicking jeep was still there, but all had vanished, leaving neither pugmark nor tire track to show where they had gone. The nature of Africa seemed to have an odd chimerical quirk, for it never let you see the same thing twice.

Back at the lodge, I said to Luke, "I feel as if we have lurched over half of Africa!" Carrying our heavy camera bags, he said over his shoulder, "If you want the *real* truth of it, sister, we clocked exactly twenty-eight miles this morning by my speedometer."

"Where else on earth could you see within a twenty-eight-mile loop an antelope's masked ball, a pair of honeymooning lions, a desert oasis like a Tiffany brooch of sapphires and emeralds and four naked bathers out of

Nowhere frolicking in a rustic spring?" was the question I wrote to myself in my diary, and rushed out to Samburu's river bar for a martini with my comrades, to build up strength for lunch.

Luke lifted his glass to his ladies and announced we could all lie up until four o'clock, when he would take us on a final game run along the riverbanks.

VI

KENYA'S
NORTHERN FRONTIER
PROVINCE

MARSABIT NATIONAL RESERVE, one hundred and sixty miles from Samburu across the stony desert of Kaisut, is as far north as Kenya tourists usually go, unless they are on a camping safari with their own tents, servants and food supplies. Most people fly up from Nanyuki, a one-hour trip as compared to the six hours we spent grinding it out in our Land Rover, thus missing many an extraordinary surface sight that was our reward in those difficult northern deserts.

The northbound track out of Samburu was without a curve for mile after mile, fairly smooth at first but soon becoming a rutted "washboard" that rattled the car so noisily we had to shout at each other to be heard. Blue domes of distant mountains grew taller and more mysterious as we approached them. Birds of prey inhabited those volcanic domes, which themselves often resembled stone bird heads above tumbled wastes of lava boulders. For an hour, I watched one tremendous dome slowly shape itself into a perfect owl head a thousand feet tall. It seemed to be looking down on us and turning its flattened face to follow our progress around its rocky base.

This was a wonderfully spectral land to explore, an unearthly bestiary of black monsters created by volcanic upheavals when the Great Rift fault was at its most active.

During a halt among the looming domes, a light plane of one of the air-safari companies passed overhead. How its altitude must reduce our intimidating domes and crags to molehills! How thankful I was that we all still had the strength and stamina for our rough surface travel over mighty Africa. After we had satisfied our thirst, I saw Luke pulling on his pigskin driving gloves as we began to climb Marsabit Mountain and knew it was a sure sign of a rough route waiting ahead.

This mountain was in fact the steep backside of a volcanic cone, and now the road resembled a jagged stone stairway mounting through forests of wild olives and mahogany trees draped with Spanish moss. Luke wrestled the Land Rover as if he had a bull by the horns, and we all clung to the handrails and rose off our seats for each punishing jolt, but carefully, for we risked bumping our skulls on the car roof. I heard him shout over the din of grinding gears, "Hang on, ladies!" as he assaulted what looked to be a wall of rock across the road that now seemed to have become an impassible elephant track.

I looked at Lou posting from her seat as though astride a bucking horse and decided that maybe I too had still a few more knee bends left in me. Presently the Land Rover paused, then made one last brave leap forward. We had reached the top. Luke now swung us off the dreadful track onto a flat, grassy clearing on the crater rim that overlooked Paradise Lake. Once again, as always happened on this trip, the punishment we had chosen to take to reach this unmapped eyrie seemed a small price to pay.

Paradise Lake, far below us, was like a sheet of glass that reflected the sky from the floor of the crater, surely the purest stretch of water that existed on our planet earth. Luke said that he had brought a small safari party up here to camp several days in this same green clearing. From the rim, then as today, the blue lake seemed to have no other purpose than to reflect the blue skies. But when we looked through our binoculars we saw the wildlife it supported — water birds on the wing (big enough in reality to be ducks and cranes) flitting and dipping over dark shapes that were really elephants ambling along the water's edge.

This was the paradisiacal lake that Martin and Osa Johnson discovered and poetically named fifty-two years ago. This was the site of their fabulous camp, where they lived for four years, photographing the family life of all the elephants nearby in the Marsabit wilderness. The sculptor Carl Akeley had come here to collect specimens for his sculptures, now in the American Museum of Natural History in New York. Once the elderly George Eastman visited the Paradise camp. For him they built a guest cottage over the elephant run and from there he could contemplate wildlife at his ease.

Luke showed us the site of the Johnson camp and said he would drive us there after lunch. There wasn't much to see of it now, for a fire some years ago had destroyed it. "But the elephants are still there," Luke said. "With luck, we might even catch a glimpse of Ahmed."

Ahmed was reputed to be the biggest and oldest elephant on earth, and he carried tusks estimated to weigh a hundred and seventy pounds each. He was a national treasure of the Republic of Kenya, preserved by a presidential decree of Mzee Jomo Kenyatta and given a special game warden, who followed his movements through the Marsabit forests to protect him from poachers. Luke had seen Ahmed only once and he hoped that we also might have that unforgettable experience. "I can't promise it," he said. "It's a matter of luck."

Before preparing our picnic on the crater rim, he inspected the forest around us, found a glade that suited him, and told us not to stray far. "Buffalo," he warned, and went back to the car to fetch the lunches he had brought from Samburu.

This forest glade was filled with dead leaves and moss from the surrounding trees, which made a spongy humus under our feet. One strange, blood-red flower pushed up from the humus and when I saw Juliet bending tensely over it, I knew she had made a botanical find of first magnitude, even more important than Ahmed and his enormous tusks.

"Look!" she cried. "Here is *Haemanthus multiflorus*! Do you realize that this blooms for only *one week* each year? *Seven* days . . . and *we* have the luck to arrive here on one of those seven days!" Her voice broke with excitement as I knelt to photograph as closely as possible this scarlet globe. It was at least six inches in diameter on a single short, stocky green stem, which thrust up from the brown humus as if by an explosion. "The amaryllis family . . . one of its rarer ornamentals. Just *look* at that inflorescence!"

In my viewfinder it looked like fiery lace, a globose

head composed of a hundred small red flowers spreading out horizontally from short, narrow tubes. "You've seen this before," Juliet went on. "On Kauai, remember those three potted bulbs I set out in my lanai every year and watch for the blooming?" I remembered. And how she invited all her friends in to see this glory.

Now Lou was also on the search and soon called out that she had found three more farther down in our glade, but alas, too deep in shadows for a photograph. Since we had promised Luke not to stray far, we climbed back to the open clearing. While we lunched, Luke went down to look at the lilies, which Juliet had pronounced worth his entire safari just to see in bloom. "Write to-day's date in your notebook," she advised. "Next time you have some very special clients like us in early September, be sure to bring them up here for a *really* rare and lovely sight."

We had an even rarer viewing on the rocky road from the rim leading down into the crater. Every forest glade on either side of the track had a carpet of blooming *Haemanthus*. Like strings of firecrackers set off simultaneously, the explosions of color continued deep into the forest's darkness and lighted it as far as our eyes could see.

Luke stopped the car often so that we might shoot these extraordinary scenes, but he refused to let us get out of the car to follow the lily paths that led beyond the curtain of forest and sunny glades. Once, made desperate no doubt by our pleading and nagging, he himself got out and with Juliet's camera he shot what looked to be a wide meadow of blooms, as visible as a bonfire through the interlacing boughs of dense trees.

"He only wants to protect us," Juliet said, forgiving him. "It's really a terrible responsibility to take three obstinate old girls like us into this wilderness. He *knows* what's out there . . . besides *Haemanthus*."

We watched how slowly and carefully he took each step and how warily he looked about him before he raised the camera to his eyes. All at once he came leaping back to us and started the motor. "I got three shots of your bloody lilies before I saw the buffalo," he said. "Here's your camera with your shots and I think you'll find the buffalo in the background of one of them."

"*Coming at you?*" Juliet asked in a shocked voice.

"I didn't wait to see!" Luke smiled to himself and kept his eyes on the perilous track leading us down to the lake.

Late in the afternoon we passed through the frontier town of Marsabit. There was a chilly drizzle. We were up more than five thousand feet and in the first really cold weather of our trip. Beyond the town, the road climbed even higher into foggy forests with trees, as always, draped with Spanish moss. Everything in the domain of Ahmed the elephant was dripping and black. In a small clearing high on a crest, Luke stopped the car beside a Land Rover with a park license and opened the doors for us.

"You go on ahead," he said. "Don't bother about your bags, someone will be coming down for them soon." He showed us the path that led uphill through long yellow grass.

Suddenly, out of the mist at the end of the path, a large German shepherd dog came bounding down to meet us. For Lou and me, it was an apparition more astounding than Ahmed would have been — and much more lovable, I might add. This handsome dog was like our own two shepherds back home on Kauai; it was as if one of them had come down to greet us. He was black and tan with a noble head, heavy coat and a strong, plumy tail, which he was wagging so vigorously that it parted the tall grass as he came through it.

His owner, the young camp manager who followed him, whistled him to heel and apologized for our reception. "He always hears the cars coming before I do and gets down here first!" Luke introduced us while Lou and I were trying to tell him not to worry about his dog annoying us, we also were "German shepherd people" and had two just like his, back home. We followed him up the hill to two green canvas tents reserved for the ladies in Luke's party. They were staked out, one behind the other, on a slope, and at first glance were the only habitations in sight, so discreetly were the other tents placed among low bushes and tall grasses that covered the crest.

The manager indicated two isolated canvas enclosures. These were the shower and toilet. At the top of the hill were the dining tent and bar terrace. "Come up and have a drink when you're ready," he said and left us with his dog to supervise the porters. Alone now, in the dripping mists of Marsabit, we felt far from our planet below.

"I'm glad our tents are close enough to hear each other shout in the night," Juliet said, not quite laughing. "This is the wildest camp I've ever been in . . . *and* the most beautiful."

The drizzle had stopped but the heavy mist still clouded the outlines of our surroundings. Between our tents and the main pavilion a few dead trees remained with black bunches of Spanish moss like funereal plumes. It was a stage set for Götterdämmerung after the gods had departed.

My Wagnerian impression was intensified when we arrived at the main pavilion above. Camp chairs were set out around an enormous fire of cedar logs. The scented smoke went up in spirals to mingle with the encircling mists, which closed us off from long views in all directions. In the circle of firelight all the activities of the sprawling camp were centered, bright and secure. The kitchen tent behind the dining pavilion was a Nibelung cave that shot out sparks to the sounds of black laughter and the clang of iron pots. The young blacks setting the tables in the dining tent tittered together in falsetto voices, like Rhinemaidens foretelling a fearful future.

Luke prepared our highballs at an impromptu bar table set up behind the family circle of camp chairs. He served us and sat down beside Juliet. I heard him ask her if everything was okay and saw her rapturous nod as she touched her glass to his in silent thanks for bringing us to this fabulous mountain. Other camp guests now appeared, mixed their own drinks, and sat down with us to stare silently at the glowing cedar logs and enjoy their private dreams.

"I have to pinch myself to realize this is Africa," I whispered to Lou. "*Oui!*" she whispered back. "*C'est fantastique!*"

We dined our first night on Marsabit Mountain more delectably than we had yet dined anywhere in Kenya. Most of the food was locally grown in the plains below the mountain. There was thick barley soup, young lamb roasted on a spit, the pinkish new potatoes we had seen for sale at crossroads on the plains, a local cauliflower and the Kenya cheeses with tinned pears for dessert.

The wide front end of the dining pavilion opened out to a wood-planked veranda under a flap of canvas, beyond which nothing could be seen in the total darkness that had fallen over the camp while we ate. Coffee was served on the veranda, away from the lights of the dining tent. We hoped that from there we would be able to see more clearly without the lamps, but we saw absolutely nothing. In the enclosing darkness you felt that if you walked ten feet away from the veranda you'd run straight into a velvet curtain and become instantly lost beyond recall.

I stepped out to look up from under the canvas flap. Not a star was visible. The blackness above was as impenetrable as the darkness below. I could feel it entering me like my breaths, spreading fear throughout my body. It was a most astonishing emotion to discover in myself, who had never from childhood been afraid of the dark. "This is the *darkest* dark I have ever seen," I told my companions at the coffee table. "It's a good thing none of us scares easy!"

"No problem there," Luke said. "I'll take all of you down to your tents. Anytime you're ready, I am. I hope you remembered to bring your big torchlights." Indeed we had. I was holding mine like a shotgun across my knees. He took Juliet's arm and they led the way downhill.

The four moons of torchlight caused strange shapes to leap out of the darkness — a whitewashed stone to mark the path, the bleached trunk of a dead tree, a clump of

75

dry, yellow grass moving slowly at its roots as if a snake were lying there. . . . The darkness of fear I had breathed in when looking for stars made me remember the puff adder that had transfixed me in the Snake House back in Nairobi — a nocturnal reptile, I recalled, with a deadly venom that acted almost instantaneously. I pulled Lou with me in a wide detour around the clump of yellow grass that was moving in no wind, and hissed, "Snake!" When we had passed it she switched her light to my face for a quick, clinical examination.

"*Ne sois pas folle!*" she scolded. "We haven't seen a snake *yet* in all of Kenya. This isn't the Congo, *mon Dieu!*" Soon our lights showed the backs of our safe, green tents.

In front they glowed with the warm light of butane lamps, which the servants had already lit. We stopped first at Juliet's tent to hear Luke's instructions about turning off the lamps and zipping up the tent flaps tightly, especially those connecting with the floor canvas, to keep out what he called "any blighter that creeps."

I lay awake most of that night, battling that nameless fear the African dark had put into me, trying in vain to find which self of mine had gone berserk and taken over control. After all lights were out, the inside of my tent became part of the utter darkness outside. I stared into a solid blackness with wide eyes that saw nothing. As with blind persons, my hearing became more acute. I listened to an insistent tapping on the tent top and imagined a myriad-footed Something walking about overhead. After an interminable time of tense nonthinking, I realized that a tropical rain had started. Then later on, the canvas wall beside my cot began to lean in against me with intermittent nudges as though some big animal were prowling out there just a few inches away from my pillow. I froze into immobility like a hunted animal and pretended it wasn't there. Thus I lay, scarcely breathing, until the gray light of dawn revealed that it was the wind that pulled in the tent wall and then released it, with that periodic rhythm I thought was the breathing of a beast outside.

"What a fool I am," I thought. Yet wiser somehow. I knew more about myself now. Never again (even in secret self-congratulatory thought that I'm ashamed to say I sometimes indulged in) would I be able to call myself an evolved being free of fear. My traumatic passage through an African night had led me deeper into my own inner darkness and had shown me the mindless animal that lived there — an indisputable part of me, dear God! a hidden "self" I never knew I had.

That it had been my first night in Africa under canvas did not explain away the shaming memory of my hours of quaking. Nor did the fact that I now knew myself to be highly, almost pathologically, vulnerable to suggestion. I should say here that I never again felt a twinge of fear while in Africa, even the next night in that same tent. Darkest Africa gave me only once the experience of its primeval power of suggestibility over the human mind. When it was over, I was grateful for this knowledge.

Looking back on it now, I think that anyone who goes to Africa and never once experiences a tremor of atavistic fear misses something very important which that ancient land can tell us. I cannot say precisely what that "something" is, but I rate it beside the awe and wonder of Africa's wildlife and the paralysis that seizes the mind when one contemplates the apparent endlessness of the African landscape.

VII

MARSABIT—WATERING
OF THE CATTLE

I'M TAKING YOU OFF the mountain this morning," Luke announced over breakfast. "Maybe by the time we get back, this nasty drizzle will be over."

"Are we going to hunt for *him*?" Juliet pointed to the large photograph of Ahmed that decorated the back wall of the dining pavilion. Luke sipped his tea thoughtfully and kept us waiting for his reply before saying, "Maybe even better than that." He beckoned to a handsome young native in baggy European suit and rumpled cotton hat who was waiting outside the pavilion. This was Farah, who was going to guide us out to the Oolanoola wells to watch the watering of the cattle. "Without him, we'd never get near the place," Luke explained. "His family owns one of the wells."

A true impresario of his safari world, Luke seldom volunteered advance information before the curtain went up. Was secrecy the occupational hazard of a safari escort? We all managed to hold back a babble of questions until we were safely down off the misty mountain and rolling again on the desert, where our driver could see any possible danger for miles in all directions.

As we advanced, Luke gave us a few headlines on the watering of the cattle, a strictly controlled activity of the desert tribes. A man's herd could be driven to the wells only twice a week, and then only to the well that belonged to his own tribal group. There were perhaps five or six wells where we were going, but we could photograph only the one that belonged to Farah's family. "No sneak shots anywhere except where and when Farah points," he warned. "Then, after he hands over the baksheesh already provided, you can shoot the works openly."

When we saw the cattle from a great distance, we thought at first we were seeing a dust storm blowing up from behind a long, low ridge. Luke steered us toward it as Farah directed. At the gravelly crest, he stopped and we looked down on an incredible river of humpbacked cattle running horns to tails down a powdery track immediately below us — running, bucking and bawling, a stream of bovine backs flowing downhill, two or three abreast, with every muzzle stretched forward toward the unseen wells which, Luke said, were still a quarter-mile away.

Upslope, running in our direction, this seeming flood of animals was visible as far as we could see and in the periodic lift of the great dust cloud, we could see other, smaller streams flowing into the mainstream from all directions of the desert. Such was our astonishment at this panorama that not one of us thought to take a picture!

Luke was watching the cataract of cattle with his experienced, measuring eye. With an electric shock of delight, I knew that he was looking for his chance to join this uncanny procession. Now and again appeared a lean native in breeches, waving a herdsman's staff to hold back the creatures behind him. Farah hailed one of these men, who helped Luke find a momentary opening. Our Land Rover made a short, growling leap down the embankment and there we were right in the middle of the bawling cattle, their rumps ahead of us, their horns behind us, and, wherever the track widened, a rush of running mates on either side of us. Indubitably this was the ride of our lives, not one of us could speak from sheer excitement. Only the veteran Farah cried, *"Mzuri, Bwana . . . mzuri"* ("good, good!") to compliment Luke on his remarkable driving. After a time, the track narrowed to a slot and the herd came to a bucking standstill with us still in the middle of it.

We could breathe freely once more, and in the pause Luke took off his sunglasses and wiped the dust from the lenses as calmly as if he had stopped at a traffic light on a city boulevard. "The wells are just ahead," he announced like a small news item.

The cattle also knew that the water was near. The mooing, bawling and bellowing from a thousand parched throats rose in a desperate ululation as they began moving slowly down the final slope. With dramatic suddenness we emerged behind them to the edge of a circular hollow. There at the bottom were the wells, in an irregular line across the desert floor that followed the meandering of the underground river supplying them. Crowds of natives were watering their herds around each well but we were too far away to see exactly how they managed this. The cattle we had run with downhill knew which well was theirs and trotted off in the direction of it, opposite to where Farah was pointing.

The whole scene was one of incomparable strangeness, a melee of seeming disorder that was operating with the precise order of some mysterious tradition unknown to us.

Luke drove slowly on and parked in a flat place just above the well of Farah's family. Farah leaped out and ran ahead to talk with an old man who appeared to be in charge. It was quite a palaver and baksheesh was passed before Farah beckoned for us to start on our way once more.

In those final steps downhill to the edge of Farah's well, we walked out of the twentieth century into scenes like those of the Old Testament. The ancient way of lifting water from a deep well up to the surface troughs was so astounding and unexpected that we had to watch it for a long time before we could begin to comprehend.

There were actually three lava-stone wells, or maybe cisterns, rising one above the other in stepback up a slant rock wall. We stood on the rim of the lowest one, built over the original spring, and looked down some twenty feet at two girls hip-deep in muddy yellow water. They wore necklaces, earrings and bracelets and their cotton saris were knotted becomingly over one shoulder, leaving the other bare. We could see the play of young muscles as each girl scooped water into a half-gallon skin bucket and passed it overhead to a man standing on the slant above. This he took from her and emptied backward into the cistern above him. There yet another man stood, scooping and passing upward to the top cistern the water that had now been lifted some forty feet up to the surface troughs, where the long lines of cattle waited.

Two teams worked upward from the girls in the lowest well — all above were men — two perpendicular lines of black bodies wet and shining, all working, like a single, coordinated unit of machinery, to a continuous chanting that beat out the staggering pace of the action. You seemed to be watching a human waterwheel at work. The lifting went on and on without respite, sometimes the metronomic chanting even quickened, as the leader stepped it up for a few rounds. Then you heard the wild laughter from the water-lifters swinging faster to the accelerated rhythm and the buckets flew upward from hand to hand as though endowed with a life of their own.

After a long study of the incredible scene, I began to take pictures. The midmorning sun, obligingly bright, allowed me to shoot at a shutter speed fast enough to stop action. Through my wide-angle lens the two dominating colors of my compositions stood in sharp contrast — black for the bodies and the lava stone, golden yellow for the water and the desert mud smoothed hard around the well rims. I knew I would come out with photographs like those old biblical illuminations in black and gold.

Luke talked Swahili with Farah, gathering the facts about this well, which he would later pass on to us. The well had been in operation for more than forty years and had been dug by Farah's grandfather, who had also constructed its remarkable earthworks. The buckets the people used were made of skin from giraffe necks, the toughest hide known, which lasts indefinitely. Only the water that flowed naturally from the spring beneath the lowest well was taken out each day and no mechanical pump would ever be introduced here, since it might deplete their pure underground source. All the desert tribes that came into Oolanoola obeyed the same rigid rule of water conservation, taking no more at one time than what the earth gave up to them naturally. And, of course, they brought here no more cattle at one time than the known number they could water.

I saw Juliet and Lou photographing the trough above the wells and walked up the path to join them. Here from the top, the entire floor of the valley could be seen and the chanting from the other wells could be heard. Knowing now what a stupendous activity was going on in each of them, the magnitude of this cattle watering boggled our minds.

The trough was a trench dug in the yellow earth, long enough to accommodate some thirty cattle at one time. Small boys sat on their heels on one side of the trough, watching the cattle lined up on the opposite side with their muzzles plunged in the muddy water, drinking with long, sobbing intakes and not raising their heads until their time was up. The boys signaled the time with a wave of slender staffs, and the line of cattle moved away single file, while another group of thirty moved in immediately.

While new herds of humpbacked cattle sprang from the thornbush that rimmed the hollow, I gazed mesmerized at a way of life that probably had not changed in a single detail since man first learned how to find water under the desert and bring it up to the surface. A few of the new herd came bawling down to Farah's trough from where the preceding drinkers were being driven away. I saw that there was even a *timing* at work in this ancient watering operation. Even the cattle, holding back or springing forward, seemed to know! This realization was one of the highest emotional moments given me by the continent of Africa.

I sat down on the boulder beside Luke and said: "No *wonder* you couldn't describe the cattle watering! It's got to be seen to be believed."

"Good show," Luke agreed, smiling. "Mattrafact, it's a *damned* good show."

By the time we were back on Marsabit Mountain the sun had burned away the drifting mists. Wherever the brilliant light broke through its dense canopy, the forest shone to its depth and luxuriance. Luke drove in slow motion toward our tents, and told us to keep our eyes peeled for a glimpse of Ahmed. But our vision was still exhausted by images of the Old Testament scenes at Oolanoola wells and at last we looked out from the Land Rover with only a desultory interest, completely drained.

A new party had arrived at the camp while we were away. Two jeep trucks bristling with sound-recording equipment were parked in the flat area, and the ground around was strewn with immense telephoto lenses, strobe lights, tripods — a fortune in photographic equipment was lying there unguarded, laid out to dry.

"Good heavens!" Juliet exclaimed. "*What* can this be but *Hollywood?*" Her guess was right.

Our young camp manager, looking years older and rather frazzled, came to tell us that eight men and girls from Hollywood had descended upon him without warning. They meant to stay for a week to make a film documentary of Ahmed under the direction of George Plimpton. They claimed to have confirmed reservations which, due to some snafu in the radio connections, had not been received here. So now he had to find camp space for them while they were out with Ahmed's askari for their first exploration of the forests.

"I've given them my tent and have moved some of my kitchen boys out of theirs for the night," he said, "but I'm still shy about three cots." He gave Luke a long, thoughtful look and then another even longer before he said, "Listen, old boy, I may have to double you up with one of them." Luke nodded reluctantly and said, "Anything for an old friend . . . as long as you don't give me someone who snores."

The Hollywood crew came back from the forests late in the afternoon while we were having our sundowners on the dining tent veranda. They were the new breed of moviemakers, young, serious and self-contained, all clad in Levi's, boots and leather vests. They disposed themselves with the casual grace of cats on a patch of dry grass outside the pavilion, leaning on their elbows and drinking lemon soda or Coca-Cola while they gazed upon the Marsabit forests below. Two girls loosed their hair prettily in the sunset light and sat with ease in the cross-legged yoga position of meditation.

We never learned which new arrival was the famous George Plimpton, because later, when we were crowded together around the tables in the dining tent, the newcomers introduced themselves by the name of their function on the team — producer, sound technician, script girl — and talked at length about their project to film the elephant Ahmed for an ABC network show, due to be shown in the United States by January 1972.

So it happened that only then, four months later, when we were back home on Kauai, we discovered George Plimpton in both film and print. *Life* magazine published his vivid piece about hunting our Ahmed so that it would coincide with the ABC documentary. Strangely, we also met Ahmed then for the first time, and it was through Plimpton's tour de force. It was he who showed us the immense, wrinkled mammoth as he moved majestically, almost mystically, from his Marsabit forest; "a gaunt, sepulchral apparition . . . of both majesty and antiquity," were the words Plimpton chose to describe this monster of nature's art and quiddity.

We left Marsabit early next morning in the usual mountain mist, which forbade our further search for this most famous of elephants. As we made the steep descent through the forest, we came upon the searchers who would not give up. They occupied the moviemakers' jeep, parked under the dripping trees, with the two shivering girls sitting on its roof, looking off into the shrouded forest where their men were soundlessly trudging. The girls were so intently listening through some device, which perhaps signaled the men's whereabouts, that they did not even look up as we rode past.

"The gods have sure got to be with them," said Juliet.

"*And* the weather," Luke added.

The mists persisted even after we were well down in the desert, and there were some lovely, clouded views of strings of camels carrying water in clay jugs across the bleak terrain. "This is as far south as camels come," Juliet told us when Luke stopped to give us a try at photographing them in zero light. Later on, we overtook some Somali ostriches, which trotted away from us with white plumes waving above their blue-gray thighs. "Males," Juliet announced. "They're the ones that carry the beautiful plumes." As their tall black bodies pranced out of camera range, I could hear the apt music Walt Disney used for his extraordinary animal films, that sedate, pacing rhythm that accompanied his ostrich scenes.

Finally we came into sunshine midway across the Kaisut Desert and peeled off bush jackets and sweaters. When we hit the paved highway leading south to Nanyuki, Luke said, "Back on the old milk run again!" and grinned at Juliet, who knew what he meant. The "milk run" was the great loop of paved highways that connected the principal game reserves — Mount Kenya, Serengeti, Ngorongoro, Manyara, Amboseli — all these the mainstream of the ordinary tourist travel.

Marsabit, in its wonderful wildness and isolation, seemed to be a thousand miles away as we rolled into the deluxe grounds of the Mount Kenya Safari Club, founded by William Holden and a group of notable big-game hunters back in the days when shooting safaris were in vogue. Powdered red with desert dust, in rumpled, mud-stained outfits, we stepped stiffly from the Land Rover and looked about at the Beautiful People while Luke went into the lobby to register his party. "Do you suppose any of them ever put a toe in the bush?" Juliet asked in a laughing undertone. Then, since she had been here before, she named the chief delights of this posh resort. "Here at last we can *all* shampoo our hair," she said happily, "and wash all the filthy things we've been carting around since Samburu. The tubs are sunken pools big as Roman baths . . . absolutely divine!"

Luke returned with the key to our cottage and Juliet called Lou and me away from an entranced examination of a wicker basket swing built for two. "Come along girls . . . *let's live it up!*"

VIII

BOTANIZING
ON MOUNT KENYA

On safari, certain domestic labors must be accomplished every so often, and only in the rare places where water is plentiful. While the larger resorts have an excellent laundry service, there remain those dozens of drip-dry necessities the bush traveler prefers to do himself, using the liquid soap he has brought along in plastic bottles.

We had been out nine days when we arrived at Mount Kenya Safari Club, and we certainly looked it — from matted hair to muddy socks. Our club accommodations, spacious and elegant, were adapted perfectly to our needs. Our private cottage had two bedrooms, which were separated by a salon with its open fireplace, and then a bar. The two bathrooms made us shout with joy. The tubs were blissful pools with steps going down to them. Smiling at our cries, the smartly uniformed servants brought in our dusty bags and we went to work.

Hair first . . . then the baths . . . then the laundry. Juliet had warned us not to forget to hang out our dresses in the bath steam and, like magic, hot water poured from the huge spigot as from a firehose.

Next came the cameras to clean and all the special lenses to be carefully dusted with a small air brush. The innumerable aluminum cartouches of exposed film must be retrieved from the corners of suitcases — and these were now emptied completely for the first time since leaving Nairobi. When you are living out of only one suitcase for a long, harried month, you discover upon emptying it items you forgot you had brought, or thought you had lost.

Two hours later Lou and I put on clean dresses and went to the salon, fit to be seen again. In our rush for a bath, we had not noticed that this highly decorated room resembled an African museum. One wall was covered by a zebra hide; another held rows of African masks carved from teakwood and an arrangement of spears and war clubs. Beside the bar hung a circular medallion, four feet in diameter, made of the black and white pelts of Colobus monkeys arranged around a star of black fur. I shivered when I touched it and moved away to the end tables, which Lou said were real African drums. The base of the standing lamp was an elephant foot, the stump cut off about eighteen inches above the four blunt, lacquered toenails.

Soon Juliet arrived, rested and radiant in a dress of royal purple with matching velvet shoes, and then Luke, in a dinner suit scrupulously in press. He pretended astonishment at our striking transformations and said with a twinkle, "Well, well . . . so you all got tarted up!" He went behind the bar and mixed drinks with elaborate ceremony.

With the first highball, I stopped trying to count the number of Colobus pelts in the furred medallion. A ridiculous activity, considering I had not yet seen a Colobus monkey and had no idea of its size. No matter, for such a rare and intricate piece of furry art must be the only one in existence. . . .

But I was due for a shock. It came soon after Luke had led us to the club's main building, with its offices, shops, dining hall and bar. There, everywhere I looked, those monkey medallions flashed black and white at me from the walls, and most of them were even bigger and more dramatically designed than the one in our cottage. Not only were there enough fur medallions to have decimated an entire Colobus colony, but there were also numberless horned heads looking down on us with shining glass eyes. The entire antelope family was represented there, from the delicate dik-dik to the giant eland, as well as every other specimen of Africa's horned ungulates.

These inhuman displays had a shocking effect on me. I tried to convince myself that such barbarities dated from colonial days, when big-game hunters roamed unchecked, that this was after all a *man's* club, created originally by men to glorify the role of the hunter . . . but I could not talk myself out of my state of anguished revulsion.

During dinner, an even larger Colobus fur medallion placed over the music stage mesmerized my gaze no matter where else I tried to look. Bound fast in my ungovernable emotions, I hardly realized my helpless state until, on leaving the dining hall through a patio hung with elephant tusks, I fell flat on my face.

My friends rushed to pick me up, all crying at once, "Did you hurt yourself?" My hand was skinned and bleeding around the base of the thumb and was the only place that hurt me. Thank God no broken bones. I let everyone believe I had been trying to read, on the engraved brass plaques beneath each elephant tusk, the names of the proud killers, but it was of course the Colobus monkeys that were the cause of my misadventure when I experienced that identification with their fate. . . .

Weeks later, in another park, when I saw my first living Colobus monkeys, I knew that I must have foreseen in some strange way the absolute beauty and innocence of these great black-and-white primates with their mantles of pure white hair draped from shoulder to tail, a scarf that floated after them through the treetops where they spent their lives — the most elegant arboreal creatures in all of Africa.

Next morning at breakfast on our private terrace, I saw the other face of the Mount Kenya Safari Club. It was as beautiful as one's idea of Eden. The acres of lawn that spread before us were the home preserves for birds of every description — red-billed ducks, Egyptian geese, storks, and a whole flock of sacred ibis feeding in the man-made pond at the foot of the slope. Three gorgeous crowned cranes came honking up to the breakfast table, begging for bread, which they took from Lou's outstretched hand while I photographed them. Never again would we be so close to these marvelous terrestrial birds, as stately as kings and crowned with a black velvet cushion that sprouted yellow feathers.

"While you're about it, you'd better get pictures of those sacred ibis," Juliet said. "Really a very ancient bird . . . you've seen them, of course, in the frescos of

Egyptian tombs. They're no longer found in Egypt, but thank heaven they have natural habitats like this one."

As I walked slowly down to the pond to photograph the white birds with naked black heads and necks, only my sore thumb reminded me of how intensely I had hated the club the night before. Now, in the morning light, the club was an avian paradise over which Mount Kenya stood guard — clear, unclouded and close enough to seem a planned part of the luxurious landscape.

We had come to the club for the sole purpose of going up Mount Kenya, for this was the excursion Juliet had always longed to make and always heavy rains had defeated her. Now, dressed in our safari outfits, we sat on our terrace watching Mount Kenya while waiting for Luke. "If we can only get up as far as the Lobelias," Juliet kept repeating while Lou and I tried to will the clouds away from the mountain. "Lobelias, Senecios, heathers and many true alpines. . . . We may find scenery like our own Alakai Swamp on Kauai. And even the same phenomena of gigantism and dwarfism."

We had all at one time or another made the strenuous hike across Kauai's high swamp named Alakai and I had even published a piece about it. "The antediluvian world of quaking bogs and stunted as well as giantlike vegetation, where violets turn into trees, trees into ground shrubs, and every sense you ever had about customary nature is turned upside down," I had written. The prospect of comparing Mount Kenya's Alpine highlands with ours on Kauai was exciting for us. The altitude in Kenya was twice as high, however, and its boggy world was probably far bigger than our own. And even if there would be nothing to compare, we would nevertheless be seeing one more of the Alpine bogs so rare on the planet that you could count them on the fingers of one hand. Save for the Hawaiian high bogs, and (Juliet thought) a possible few in the Andes, most of these highland habitats of prehistoric plant life were in the equatorial mountains of Africa — in Uganda's Ruwenzori, which Stanley named "The Mountains of the Moon," on Mount Elgon on the borders of eastern Uganda and Kenya, and here on Mount Kenya above the ten-thousand-foot level.

Luke fetched us at nine o'clock in the freshly washed Land Rover in which he had already packed our four picnic lunches, which meant we could hope for a full day. He said that the weather reports for the mountain were "promising as far as you can trust these native blokes," and that he had decided to make the ascent via the Sirimon Track. This was our unique adventure of a "plant run" instead of the usual "game run," an interlude that was to give us — through a spectral vegetation that had preceded all other terrestrial life — our most thrilling perspective on Africa's unimaginable antiquity.

The Sirimon Track branched off from the paved highway a few miles north of Nanyuki, a rough scar of red earth that mounted straight up toward a high-altitude forest. The national parks guidebook, which I tried to read between jolts, assured the visitor to Mount Kenya that he need not be an experienced mountaineer and if he had a four-wheel-drive vehicle, the access to the Alpine moorlands was easy.

Nevertheless, the advice of the African guardian in charge of the park gate was more practical. He gave us a typed bulletin with the title "Mountain Safety" to remind us that there had been fifteen deaths on Mount Kenya. There were complete instructions on where to find stretchers, oxygen sets and medicines for all types of casualties. "Be sensitive to your own health," the bulletin advised, "and that of your companions. Do not go up the mountain unless you are in good health. If anyone falls sick, GET HIM OR HER DOWN OFF THE MOUNTAIN AT ONCE, LOSE HEIGHT QUICKLY."

Luke signed the book, giving the approximate afternoon hour when we would be due back at the gate again, and the guardian saluted us smartly as he raised the barricade.

At once the track became rougher and steeper, cut by runnels of rainwater, which Luke stopped to study before deciding which ones he could safely set his wheels into. We gave him the support of total silence as he ground upward through the high forest and into the zone of bamboos so densely packed together that you could not imagine even a Mount Kenya mole rat getting through such a thicket. Beyond the bamboos we came into sunlight again in a region of hypericum scrub. Now we could feel on our cheeks the icy air of Kenya's twin peaks, though they were still unseen. The danger could be ahead of us over the last slippery rise of track and there was treachery felt in a deep mud puddle in the middle of our path.

"We're not going to let *that* bloody thing stop us!" Luke said. He got out, found a stick, and tested its depth. "Aha!" he said and came back to the car. With a determined face, he put gears in four-wheel drive and slowly ground through the puddle with spraying mud-like brown wings sprouting from our wheels. And so over the hump we came into the moorlands and the sight of Kenya's icy peaks dead ahead, miles away but clear as glass.

Also, dead ahead, cutting the track neatly in two, lay a body of water like a pond. Luke drove slowly toward it, cursing under his breath, and stopped on the down-side shoreline. He used his measuring stick again and when we saw it go to a depth of more than three feet, Juliet cried, "This is it, my friends! Again I'm *not* going to get up to the Lobelias! I know that look on Luke's face...."

In the end, it made little difference. We had reached the wonders of the Alpine moorland and were completely enchanted. As always, when moved by some new marvel, we strayed blindly apart from each other, Lou, Juliet and myself, each seeking her private inner experience in a strange and baffling world. For me, the dry-grass tussocks were the first marvel; much bigger than our Alakai tussocks, they looked like gigantic yellow wigs dropped over the Alpine moors as far as eye could see. Lou, waist deep among the tussocks, appeared to be meditating over some small thing she had found growing between them, while Juliet, leaning on a tall staff, stood like a statue in the meadow across the road, staring down at a plant between her feet.

Our silent preoccupations gave Luke something more to think about than the alluvial pond that had stopped him. He came after his charges and warned each in turn not to make any quick movements, but to walk slow and easy since we were above ten thousand feet. He took up a position where he could see all three of us, lit a

100

cigarette, and watched with concern the slow-motion rovings of his separated flock. He could not understand why each of us had to go off in a different direction and why there was no communication or even glad cries among us. This was not the moment to tell him that it was such an absolute freedom for each of us that had bound three such disparate persons into our unique harmonious trio.

I began to see, due to my camera, how Africa had diminished our physical forms because of the wide horizons. My comrades looked strangely small against the tremendous backdrop of Kenya's highlands. Even my long lens could not bring their bodies up to a size that would seem natural against the giant heathers behind them.

The altitude made it difficult to breathe when I handheld my camera for the long views without a tripod, which was impossible to set up on the uneven boggy ground. I decided to try first for the details of this icy immensity — the great hairy tussocks, the clumps of yellow flowering plants that resembled daisies, a tiny, mauve, crocuslike bloom that grew close to the ground. I discovered I was more stable kneeling than crouching, and changed my long lens for a 50 mm, which brought me to within two feet of the ground cover. When I began to crawl over the matted turf I discovered plants shaped like stars with flowers so small that they couldn't be seen when I was standing. This was the original carpet of this garden of the gods and I almost fainted for joy when it sprang up into focus — miniscule stars on a bed of prehistoric lichen . . . I sat back abruptly on a hairy tussock to catch my breath.

Ahead, in a kind of mossy *allée* between tussocks, I saw seven yellow straw balls lying on the ground. What could these mysterious objects be? I called Luke and he rushed to me, glad that someone in his party had begun to speak again.

"What in God's name could those be?" I asked.

He gave me his enchanting grin and went forward to kick one of the straw balls apart with his boot. He came back, patted my shoulder, and said, "That's one of the reasons I don't want any of you to wander about. Those, my dear, are old elephant droppings!"

"Elephant droppings? How on earth do elephants get up *here?"* I cried, but he was walking away, laughing to himself like a big white hunter who had just told a greenhorn one of the facts of elephant life.

I photographed the spoor, taking great care to get all seven straw balls into my frame. Though I had read somewhere that elephants eat up to four hundred pounds of vegetable matter per day, the size of the droppings nevertheless amazed me. I picked one up and found that the fingers of both hands could not meet around its great circumference.

After an hour or two Luke rounded us up in the high, thin air and we started down the mountain in search of a place for a picnic lunch. He stopped willingly to let us see the things we had missed on the way up, when our eyes had not left the slippery track. There were the hypericum trees just below the moorlands, sunbirds flying among their golden blossoms; the bamboo forest below the hypericum scrub; then the magnificent stands of juniper and podycarpus trees, in reality the "forest primeval," although on the opposite side of the globe from Longfellow's "murmuring pines and . . . hemlocks." Luke stopped a long time in the great forest, and finally selected the trunk of a fallen giant, suitably upholstered in brilliant green moss, for our picnic bench. He said that if we would sit quietly, we might see a bushbuck, or the secretive bongo, largest of the forest antelopes and the most brightly colored — chestnut red with black and white striped legs. So, as always, we ate in silence while we gazed expectantly through the dark spaces.

After our lunch in the forest darkness we drove down into sunshine and got out to warm ourselves from the hours of chills and tension on the high mountain flank. Here we had the longest view we had yet seen in Kenya. The great panorama spread out below us, brightly colored on the near side with lakes and woodlands clearly outlined before they faded off toward the west into a dazzling shimmer without details. "It's a bit of the Great Rift Valley," Luke said. "We'll be heading down there tomorrow for a look at Nakuru's flamingos."

Seeing your itinerary laid out like that in nature itself, a diorama of real woodlands, inhabited villages and the hazy soda flats of the Great Rift, you wondered if you could contain all these excessive impressions that East Africa sent forth with such reckless extravagance. Only the cameras could take and contain it all and later give it back in pictures more accurate than the fallible human memory could retain.

IX

NAKURU
AND THE FLAMINGOS

W<small>E CAME DOWN</small> from Mount Kenya's highlands via Thomson's Falls, where we stopped for a midmorning cup of tea at the inn, to brace us for the final descent to Nakuru. After the falls, the paved road became a winding, stony cart track, heading west through rugged hill country. Somewhere along this track, we came out on the summit of an escarpment and here Luke stopped the car that we might look over the forty miles of hazy Rift Valley, patched with soda lakes.

Though we knew already that this marvelous panorama was only a small fraction of the great geological fault that divides the African continent from Ethiopia's Red Sea to Rhodesia's Zambesi River, the sight of it was more than enough to paralyze our minds. The cataclysm that produced this gigantic fault in the earth's crust was not to be imagined. On the map, chains of lakes marked its course through Kenya and Tanzania and the peaks of Kilimanjaro and Elgon (believed by geologists to have been pushed up during the Rift upheavals) were the volcanic signposts along the way of this prehistoric agitation. Speechless, we looked down from the escarpment. I felt a strange excitement.

Everything we were to see henceforth lay down there in that vastest fissure on earth — the game reserves of Serengeti, Ngorongoro, Manyara and Amboseli, and the famous archeological sites of Olduvai Gorge, where bipedal mammals had made the first effort of hominoid life on the planet. This is where it all began, I thought, here is the cradle of everything that breathes today.

Nakuru seemed surprisingly modern after studying the map of Genesis from the escarpment. Called "capital of the Rift Valley," Nakuru is Kenya's third largest town (after Nairobi and Mombasa) and it has a beguiling colonial air about its wide main street lined with the white buildings of banks, hotels, bookshops and farmers' cooperative stores. Despite its contemporary appearance, Nakuru is nevertheless the ideal stopping place for one's first experiencing of the peculiar specialized life of the Rift Valley. Ideal, because here one sees mainly birds, the fabulous *Phoenicopteridae* — the pink flamingos that have made Nakuru famous throughout the world. Here, I believe, is the only place in East Africa where the satiated tourist is given only one living wonder at a time to marvel over.

I began to marvel even before we drove down to the lake. In the lobby lounge of the Stag's Head Hotel, while waiting for Luke to register us, I stared into a glass case filled with pink and red feather corsages that deftly imitated roses, double hibiscus, gardenias, dahlias and baby orchids. I picked up a folder and read about these exquisite "flowers," which are made of flamingo feathers. This was a cottage industry organized by the World

Gospel Mission in the Bethany Bookshop that had been given the exclusive right to pick up discarded feathers along the shore of Lake Nakuru and had trained the African artisans in making "flowers" of them. In return for the privilege of collecting these feathers, the bookshop made a quarterly remittance to the national parks' fund for wildlife protection and devoted the remainder of its profits to philanthropic missionary work.

"What a marvelous collaboration between birds and missionaries!" I said to Juliet. "I'd love to see the main collection in the bookshop."

"Better see the flamingos first," Juliet suggested, as if she knew that the sight of one million flamingos on Lake Nakuru would cure me of my infatuation with those feather "flowers."

We did not go down to the lake until very late in the afternoon, after too big a lunch and too long a siesta. A declining sun made good photography questionable, thus freeing us from all other activity except looking.

I could not believe my eyes when we came out upon the lakeshore. The feeding flamingos made a pink blanket covering the offshore water so completely in all directions that one could not see the color of that water except by looking far out beyond the muddy edges toward the center of the lake. There the water shone dark green and revealed its densities of microscopic algae and diatoms. On the opposite shore of this four-by-six-mile lake other congregations of feeding flamingos were seen as a pink scum that framed the water.

Light from the setting sun fanned out from beneath the cloud cover and threw a golden nimbus over the scene. The continuous gabble of feeding birds was like the sound of a humming dynamo that magnetized the senses. I walked across a black, sandy mud flat to get near enough to see just one bird in the mass of feeding flamingos.

A few dozen took to the air at my cautious approach, running over the water to gain momentum for their flight, and when they lifted off the surface they carried an upsweep of water with them that looked like fluid jade. In a 1970 *National Geographic* article we had brought along for safari reading there was the statement that the million flamingos consume more than two hundred tons a day of the mineral and algae nutrients that enrich the contents of Nakuru's soda lake. The author, Dr. M. Philip Kahl, said that the nutrients were so abundant that he could *hear* them by putting his ear to the water. "It sounded like the fizzing of a gigantic glass of champagne."

When the fading light began to dim the spectacle, Luke called us back to the Land Rover. "You'll have all tomorrow morning down here with your cameras," he promised.

Back in the hotel, I didn't look again at the lighted showcase of flamingo feather "flowers." After seeing the blooming of a million live flamingos, how could one?

Next morning we drove to the lake at eight-thirty and stayed along its shores until noon. In full sunlight the flamingos were even more bewildering than they had been the evening before. Now the massed pinks, whites and coral reds of their plumage, against the emerald water and pale blue sky, made a smashing color extravaganza that took one's breath away, as one gazed at this composition admirably and delicately framed by the violet escarpments of the distant Rift.

I recalled what I had read the night before in the Nakuru park booklet — a quote from the American bird authority and writer, Dr. Roger Tory Peterson, who had called this scene "unquestionably the greatest ornithological spectacle on earth." It was indeed — a spectacle like a huge ballet of a million winged performers on a watery stage. It was impossible to see and absorb all at once — an assault on the senses of sight, smell and hearing, a crash of color, a stench of mud flat and an unearthly babble from the feeding birds. Luckily, we had done our homework in the park booklet the night before, reading up on all the known facts about Nakuru's flamingos.

They have flourished throughout the ages because their natural habitat — the soda lakes of the Rift — is unusable by man. The high concentration of soda salts not only renders the water undrinkable and useless as crop irrigation, but it also produces serious burns on the feet of any fisherman wading into it; moreover, only a very small fish with a high soda tolerance can live in these Rift lakes.

Nakuru's flamingos do not breed here (no one knows why) but they come here to feed in mass migrations at unpredictable intervals. Since all the Rift soda lakes are endlessly rich in algae, no one has yet learned why a million flamingos suddenly decide to change lakes. All that is known is that they generally fly at night. Dr. Kahl reported that he had often heard them passing over his Nakuru home, "gabbling to each other like geese."

The flamingos' only natural enemies are the fish hawk and the marabou stork, which prey on the young chicks,

111

especially those that develop crippling shackles of salt on their legs from trotting about on the drying soda flats around their midlake nests of mud; as they grow older they learn, like their parents, to wade over to the shore-line sections of fresher water to dissolve their disabling soda shackles.

Africans have seldom been known to use their flamingos as food, although the ancient Romans are said to have relished flamingo tongues as a delicacy.

Scientists have estimated that a flamingo that survives its harsh environment can live to the ripe old age of eighty.

The flamingo feeds with its head upside down, and its bill is accounted one of nature's marvels. It is lined on the inside with thousands of minute platelets against which the tongue, working like a suction pump, draws water from which the platelets filter the microscopic algae that is the sole sustenance of these great birds.

Thus the facts, or better the program notes, to introduce the greatest bird show on earth. . . .

The foreground mass of flamingos busily filtering food with their beaks turned upside down in the shallow water looked at first glance to be a continuous low cover of pink cloud upheld by a forest of thin pink stilt-legs thirty inches tall. The unearthly babble of the feeders combined with the peculiar stench of the algae broth on which they fed attacked one's auditory and olfactory senses while the visual was being ravished with pure beauty. Through the middle of the feeding mass, bi-secting it end to end for hundreds of yards, strutted a parade of flamingos, heads up and all facing in the same direction. These, Luke said, were the courting flamingos working themselves up to a breeding frenzy, honking, babbling and shoving, with uplifted beaks. This manifestation, taking place miles away from the lake where they would eventually breed, could go on intermittently for weeks and even months. As if attached to a conveyor belt, the closely packed courting ranks moved in unison straight through the feeding birds to a certain point in the lake, where abruptly they made an about-face, all at exactly the same instant, and paraded back again.

Beyond the massed flamingos, out toward the center of the lake, flotillas of great white pelicans were fishing. Here again a strange fandango was going on. The pelicans fished in groups of a dozen or so, formed into horseshoe or V patterns. They sailed along as a unit in the same direction, with great, pouched beaks upheld, then, suddenly, dipped simultaneously into the middle of the water space enclosed by their bodies. A fraction of a minute they remained tilted down, then bobbed up to the surface with naked pouches full, swallowed their catch, and sailed on in unbroken formations until the next simultaneous dip, and the next and the next. . . .

How in God's name were they given those mysterious signals that produced this magical simultaneity of action? The courting flamingos marched, then all together made their about-face and marched back. The pelicans sailed in formations, dipped all together, then sailed on. . . . You could *see* the result of a command you could not hear. Even through binoculars you could not detect a leader signaling the moment for a mass about-face or a fishing dip. It was as baffling to watch as a magician's stage act.

We three were alone on the lakeshore for the first hour, trying to catch something in our lenses that might reveal the invisible sorcery at work among the massed birds. Then a minibus arrived on the scene and discharged its tourists. They ran at once to the water's edge, clapping and calling, to send the flamingos up in the air for the kind of unrealistic photograph television nature shows always give — a pink cloud in flight against the blue. Even the courting flamingos broke ranks and took to the air at the frightening disturbance, and soon there was no more magic to watch.

Luke drove us far down the mud-caked shore, past dead trees filled with black cormorants, to another mass of feeding flamingos, which covered the water right up to the cliff escarpment that sealed off this end of the shallow lake. Here under the cliff, the bird babble was a continuous percussive vibration of such force that it soon became unbearable. We sat in the car like deaf-mutes gazing at those acres of flamingos, unable finally to focus on a single one. I made a few shots with my wide-angle lens from the car window and then gave up in despair. Juliet spoke for all of us, saying, "I've had enough if you have. . . ." Luke drove us off through the reeds around the escarpment and we returned to town by a different route. Even as we rounded the foot of the cliff, we could hear in decrescendo the dynamic babbling of the flamingos.

Then, in the quiet of a wide grassland beyond the reeds, Africa gave us a stunning encounter with one of her rarest and shyest of animals, the reedbuck. It was then we discovered that our flamingo-dazzled eyes could still focus on single creatures. Luke stopped the car and whispered, "Bohor reedbuck." Not more than thirty feet from us, we saw a family group of those elegant, small antelopes, two males with short horns sweeping back then hooking forward, and two hornless does. Their reddish fawn coats made them look like copper sculptures as they stood motionless in the place where they had been cast. They studied us intently while we photographed them in their statuesque immobility, which seemed like deliberate posing. Then one of them sent out a shrill whistle and they all bolted off toward the reeds, running together with a delightful rocking-horse gait.

After the mind-numbing simultaneous activities of the flamingos, the sight of that graceful group of reedbuck, which live in small herds of no more than three or four, restored our reason. Thus, I thought, does Africa reveal her spectrum of wonders — from superabundance to rarity and both, so to speak, just around the corner from each other.

We had our first meal in a native restaurant in Nakuru. This was an experience far off the beaten track, since most tourists take their meals in the hotels, where British cooking is comfortably familiar. Luke had promised us one good curry before the safari would end and said that Nakuru was the place for it.

An African safari, I should explain, gives the traveler very little contact with the Africans in their own milieu. Except for hotel servants and the occasional game warden who rides with one through the reserves, the natives seldom have contact with the safari crowds continually swarming through their countries to see what wildlife is like.

I remember one night over sundowners in a game lodge when we had been out for perhaps a fortnight, I said to my companions: "You know, we've been more than a week in Africa and I have absolutely *no* feeling that I'm in a black republic." Luke looked at me with pity for my naiveté and said, "You know why? Because you're on a magic carpet that takes you over their heads. That's what a well organized safari is. . . ."

Magic carpet or not, I kept to my belief that because we lived in Hawaii, where the majority is dark-skinned, we were less aware of racial color than the average mainland tourist unaccustomed, as we were, to living in a population of black, brown, yellow and white skins.

Luke led us to the promised restaurant, which was owned, run, and patronized by Africans, an upstairs place in the main street. After our exhausting morning with the flamingos, we would not have had the strength to climb these dingy stairs had it not been for the seductive curry odors coming down to meet us. We went directly into a bright, old-fashioned dining room with paper covers and artificial flowers on the tables, and all tables but one the barman pointed out to us occupied by blacks.

Luke gave our order in Swahili to a tall Kikuyu waiter who could have been the handsome symbol of Kenya's successful Africanization program. Though no African patron wasted a glance on us, we were all eyes looking at them. Lou's reflective expression suggested that she was remembering the "boys" she had taught in the Congo just one generation earlier, young savages who called her "Mama" while learning to change dressings under her direction in a convent hospital. The Africans around us, wearing shirts with ties and neat cotton business suits, might have been those same bushboys grown up now. They were obviously men of substance in the community and pupils of the white man's ways. I fancied I could hear Lou's unspoken comment as she looked about her: "I hope my Congolese have come along as far as you have."

The waiter presently brought us foaming cold beer, set great bowls of rice and lamb curry before us, and we served ourselves family style from the center of the table. Next came the condiment tray and a long index finger pointed to the grated coconut, the mango chutney, the chopped peanuts, our handsome Kikuyu naming each condiment in turn in perfect English while he extolled the merits of his country's produce with chauvinistic pride in his snapping eyes. The curry was the spiciest and hottest I had ever enjoyed, even better than that perfect one we had in Bombay on the way out to Africa.

We ate in silence, helped ourselves to seconds, and joyfully nodded when Luke pointed out our empty glasses to the waiter. I noticed that ours was the only silent table in the place, all about us were the rumble of deep voices and the explosions of laughter. The sight of all those dark faces with their mysterious communication gave me a feeling of being left out of something I knew I would have enjoyed.

We were there where we had wished to be — among the Africans on equal footing for once, as clients in a public place, served the same food and drink. Nevertheless, inevitably, we were separated from them, not because we were tourists and white but rather for something that had to do with our differing racial developments, a kind of generation gap that made us seem old and burdened and careworn while they were the true children of the human race.

There was a distinctive atmosphere, almost like an aura, that surrounded them whenever any two, or four, or more came together. A certain atmosphere built itself slowly up among them on gusts of laughter accompanied by swift motions of mobile black hands. It surrounded them, this transparent aura of gaiety that our eager eyes could penetrate, but not in any way understand.

X

"ANY TIME IS TEA TIME"

WE BROKE THE LONG DRIVE from Nakuru to Kee-korok in the Rift Valley with a one-night stop in Kericho, a tea plantation town at the foot of the Mau Escarpment. Our late curry lunch made this an easy half-day ride and we arrived at the Tea Hotel in Kericho in time for five o'clock tea. Since none of us had ever seen how tea is grown and processed, we were enthusiastic to see another facet of Kenya's numberless attractions, which are not all wildlife by a long shot.

The Tea Hotel, a large, stone edifice, was sedate to the point of stodginess, its colonial atmosphere reminiscent of our old plantation-town hotels in Hawaii. The servants were Africans and spoke with British accents, and the lounge steward looked at us with disdain for having come to tea in our safari clothes. He led us to a table suitably distant from the dressy ladies on the chintz-covered sofas of the lounge, as if to protect them from the sight of us. It was a new experience to be put in our place by this black pillar of empire. He brought tea and cakes for us and then stalked off without a word or bow. Juliet said, laughing, "That'll teach us!"

121

We waited for Luke to unload our bags in the stone cottages reserved for us, far out in the splendid hotel gardens, and presently he came to fetch us from our sequestered corner. As he led us past the hotel desk, I snatched up a green booklet entitled "The Story of Tea in East Africa."

Two hours later we passed that desk again. The African clerks did not recognize us and Luke had to give our cottage numbers before we were ushered, this time with ceremony, into the hotel restaurant, where we now blended harmoniously with the Beautiful People who "dress for dinner." My diary notes that we had roast pork and hot apple tarts for dinner, but I remember nothing of that. I was thinking solely about the surprising facts I had just found in the tea booklet.

Tea is one of East Africa's most valuable exports. More than two million pounds of tea come each year from Kenya, Uganda and Tanzania, a cash crop to the world's markets, graded and blended as English Breakfast, Orange Pekoe, Pekoe Fannings and so forth. The major problem is to coax or even force the Africans themselves to drink tea, which will increase the local consumption. To politely promote tea drinking in Kenya, the Associated Tea Growers hold an annual "Tea Week," in which a "Miss Tea" is elected to preside and charm. Even a slogan has been chosen: "Any Time Is Tea Time." Tea is fitting for all occasions, and ours had been ordered for breakfast the next day at 6:30 A.M., before we would start out on the long drive to Keekorok in the Mara Masai Game Reserve.

Tea at 6:30 A.M. always preceded a long and busy day. Africa's far from charming practice of waking up its safari travelers with early morning tea undoubtedly keeps the safaris moving on their inexorable schedules. Every morning since we had been in Africa, that pot of strong black tea had put us on our feet at whatever hour our escort had ordered. Whether we had slept in tents or in luxurious game lodges, that tea was always brought on the dot, and it stood one up like an alarm clock. Moreover, the tea steward did not simply set his tray down and trot off. He gave a hearty *"jambo!"* and waited until someone answered with a somnolent groan. The tea boys no doubt were the real right arm of the safari escort.

Luke had persuaded the plantation public relations manager to lead us through his own domain. He came for us after breakfast and turned out to be one of the most educated young Africans we had ever met. His white business shirt set off the lively intelligence of his black face under its high-domed forehead, and his English was impeccable. He had brought a Volkswagen for Luke to drive, since the plantation roads were too narrow for the Land Rover, and drove ahead of us in his own Volkswagen.

We followed his car through the pretty, landscaped streets of Kericho to a turnoff that led us directly onto the plantation grounds. Juliet, our botanist, told us that tea was a camellia and that its Latin name, *Camellia sinensis*, was reminiscent of its antiquity and Chinese origin.

Our guide stopped at the top of an incline and gave us our first view of tea under cultivation. It was a ravishing sight of low, dense bushes with all leaves shining, all barbered off flat on the tops as if a gigantic razor had leveled them across the entire immense valley. Our guide walked back to tell us that what we were looking at was an acreage that had been recently pruned to encourage the low-spreading frame of the tea bushes. Soon again it would be ready for plucking. Only the tender top, two or three leaves with the terminal bud, was taken. "Like this," he said, and pinched off two pale green leaves and bud from a bush to show us. "One bush will produce about two pounds of green leaf a year," he said. "Now we'll go on to where they are plucking, which is the most important of all the field operations."

As we approached, we saw that the pickers were walking along narrow paths between the bushes, plucking with thumb and forefinger as they went only the delicate terminal leaves, which they tossed over their shoulders into round wicker baskets. They wore bright yellow or red plastic aprons to protect them from the water soaking the bushes. "We need rain every night to produce good tea, and we get it here," said the guide.

We asked him what was the function of the long slender sticks the pickers were waving ahead of them over the bushes. Those, he explained, enabled them to see more easily the tender young terminals growing up from the "plucking table." I watched a girl in a red apron working her wand over the bushes ahead of her with one hand and plucking with the other — but only here and there — a few leaves so small that I could hardly see them when she tossed them back into her basket. How many thousands of leaves must she pluck to fill that round basket bigger than her torso? How many tens of thousands? She moved with slow grace along her hidden path, her eyes intently on the bushes ahead while she chattered, without looking up, with a man who was plucking in an adjacent row — taller and with a longer arm reach than hers. Now and then they laughed together and again I wondered what was the mystery of that strange African gaiety.

Our guide now gave permission to take pictures and when the pluckers saw us aiming cameras their way, some of them stepped out from the bushes and let us see at last their thin black legs, hidden until now, and their sandaled feet. Farther along we came upon a weighing station, where the pluckers brought their full baskets and received a chit with the weight of their picks scribbled on it. Most of the crowd already lined up for their turn at the scales were laughing and gesticulating, but a few old men and women were sitting beside their baskets, faces blank with exhaustion.

Over a hill beyond the weighing station stood the great factory, surrounded by a cloud of perfumed air that made one almost drunk with delight. How to describe the quintessential fragrance of tender young tea leaves being processed? We followed our man into the plant, up a flight of cement stairs. Beside us a conveyor belt was carrying up the fresh-plucked tea, to be dumped into the withering troughs. Our guide explained the first step in the many processes. "We wither it overnight first, to remove the surplus moisture — about one third of its weight. The next process is to pass the withered leaves through stainless steel rollers, which rupture and macerate the leaves to release their flavor. Next comes the fermentation vats." These we saw bristling with thermometers and timing devices. Under strictly controlled temperature and humidity, the leaves now undergo the chemical changes that give the characteristic tea flavor. "We ferment no more than three hours, frequently less," continued our guide, picking up a handful of the fermenting leaves to show us the color change occurring here — from green to a copper tone.

After the fermentation comes the drying and the final sorting, when the leaves are passed through meshes of various sizes to produce the different grades of tea. From any spot in the immense building one can see the entire processing in a continuing action — a marvel of exactitude, every step precisely timed.

The green-smocked Africans standing over the withering vats, rollers and fermentation troughs kept their eyes on the processing tea, their serious, dark faces revealing a concentration deep and almost personal in character, as if they themselves were experiencing what was happening to the tea. This was another face of Africa, the face of knowledge and expertise. A pity there was not light enough to photograph one of those beautiful, intent faces, so that later I could place such a snapshot beside a picture of, say, a Samburu warrior dancer or one of the primitive natives watering his cattle at Marsabit. Perhaps I could have been able to compare in one glance both ends of the spectrum of Africa's variety....

Before leaving the factory, we passed through a laboratory presided over by the tea taster. He was a tall, handsome fellow who spent his time brewing samples of tea brought to him from the factory outside, from different steps in the processing. He had rows of glasses on his desk, teas of every color from green through golds to black, and he insisted on preparing for each of us the tea of our choice, in porcelain cups kept for visitors. His eyes, as brown and liquid as strong breakfast tea, sparkled with pride as he watched us sipping with pleasure the tea he had blended.

"The blending," explained our guide, "is the key process. This man knows all grades of tea and blends them by taste."

I restrained myself from sneaking a snapshot of the tea taster bent over his fresh-brewed samples, his wise face lively with his special knowledge that had taken a lifetime to acquire. Here was yet another face of Africa I must remember, without a picture to help me.

Strangely enough, that very same afternoon, when we were a hundred miles away from Kericho, I was able to see this tea taster's face vividly in my memory. We had made the long drive south from Kericho via Sotik to an escarpment that overlooked the Mara Masai Game Reserve, where Keekorok Lodge is located. For more than two hours we had not seen a living creature, man or beast, and when Luke drove off the rocky track and began to weave the car through clumps of thornbush, we realized we were in the wildest and emptiest place we had ever seen. Luke soon found the shade and flatness he wanted, and also a glorious view for his stiff and famished passengers of the Masailand plains far below in a white-hot shimmer of haze and dust.

An eerie silence brooded over this high, dry emptiness chosen for our lunch. When you put your attention on it, the silence seemed to become faintly audible, like a hum of high wires from above. Our striped camp chairs and the dark green hood of the Land Rover nosing into our picnic circle were the only colors in the desolation. Alkali dust as fine as talcum lay over thornbush and parched grass clumps, and we whispered to each other as if in the presence of death.

Suddenly, just as we were finishing, three Masai with long bows and knives appeared like lean, brown jinns, materialized out of the air. They strode down the dusty slope before us and stood beside the car. To our relief, they were smiling and nodding. Luke announced in a voice he made indifferent, amused, even unsurprised, "We have visitors!" He went to meet and shake hands with them. Since he did not speak Masai and they did not understand Swahili, he invited them by gestures to help themselves to anything they wished that we had left from our lunch.

The Masai, who live almost exclusively on milk, and occasionally cattle blood, looked carefully into our Tea Hotel boxes of leftover chicken legs, cakes and fruits, then shook their heads. Whereupon wise Luke called out in a loud voice, *"Maji? . . . maji?"* and went to the car to lift out the jerrican of water he always carried. This was what they wanted. They laughed at our plastic "glasses" and one of them produced a big gourd from which they would drink — a quart at a time, as Luke filled and refilled it while they passed it from hand to hand. They drank with the kind of sobbing intake I had heard at the cattle watering and when one of them emptied the gourd and looked up with water running down his chin, the other two laughed wildly at this extraordinary display of waste.

When they had finished, Luke folded up the camp chairs and stowed them in the car, signaling us with his eyes to get going. The Masai promptly squatted on their heels in the patch of shade we vacated and watched our departure with round eyes. There was not a sound from any of them. The totality of their attention and silence was oddly unnerving. We had read much about the Masai and their fearsome reputation as nineteenth-century warriors who had kept all Europeans out of their vast territory until Joseph Thomson went across it in 1883, achieving with courage and improvisation what neither Stanley nor Livingstone had been able to do. Perhaps that rather terrifying historical background produced our apprehensions as we continued to stare at the first Masai we had encountered on the safari.

What none of our reading had prepared us for was the *look* of them. They were not black like the predominantly Negro tribes of Kenya, but a shade of coppery red-brown, not unlike our American Indians. These three Masai now watching us in steady silence could have been cousins of America's Apache, Sioux or Cherokee tribes, except that they were taller and decidedly more handsome of countenance.

As I gazed at them seated beneath our thorn tree, with headbands and necklaces gleaming against their coppery skins, their bone earplugs in their enormously stretched earlobes hanging almost to bare shoulders, they made a picture that almost drove me out of my mind, such was my desire to photograph them. Luckily I knew that the mere sight of a camera aimed their way would bring real trouble for Luke, whose sole aim now, proclaimed by his anxious scowl, was to get us away as quickly as possible. He left the four picnic boxes with the Masai, raised his palm in a farewell salute, and hurried back to lock the car's rear door. Meantime we warily watched one of the warriors take an orange from a box and peel it in a continuous spiral. His knife was at least fourteen inches long.

When they saw our driver getting in behind the steering wheel, one of them came swiftly to the car. He looked through the open window on my side, staring directly into my eyes, and made some odd wagging gestures with his upraised finger, as if trying to tell me he had a secret to be shared with me alone. Luke spoke quietly: "Let him go on — see what he wants." The Masai's eyes, as round as a lion's and with yellow glints, never left my face as he pulled aside his orange cape and revealed a small leather bag hanging over his bare chest. He lifted the bag, shook it at me, and showed me its silver padlock. He reached into a fold of his cape and brought forth a key. This he held up before my face, nodding while his eyes still held mine. I managed to produce a smile. Then he opened the padlock and lifted the flap on the bag. In slow motion then, his hypnotic eyes lighting for what he was about to reveal, he put his long fingers into the bag for a second or two while searching my face. Then, abruptly ending his pantomime, he plucked from the bag a Masai choker of colored beads strung on copper wire and thrust it at me through the window. Stepping back, he waited for my admiration and thanks.

The choker lay curled on my palm. I stared at it, frightened by his unknowable impulse. All at once I saw, very clearly and reassuringly, the face of our Kericho tea taster framed in the circle of the choker, a knowable face that spoke a language of smiles I could understand. Here was the opposite end of Africa's broad human spectrum. . . . Still I felt the peculiar possessiveness of the Masai's eyes upon me and would not look up. I continued to make a pretense of examining the choker, which was indeed beautiful and quite old.

"I can't accept this," I said to Luke.

"You have to," he replied in a low voice. "These Masai are fiercely proud. You can't want to *insult* him by handing it back?"

"Could I then . . . perhaps . . . offer him something in return?"

"No, not you!" Luke said quickly. "But maybe I can. . . ." He took two shillings from his pocket and held them out to the Masai with a broad smile of gratitude. The warrior, thank heaven, accepted his offer with a courtly bow and stepped back while Luke started the motor with a roar. We slowly drove up the dusty pathway between the same thornbushes out of which the Masai had so mysteriously appeared.

I looked back once. The warrior was gazing after us, motionless. His cotton toga blew back against his body and revealed its triangular proportions from broad shoulders that tapered down to a slender waist. His was a Pharaonic figure, leanly sculptured by the wind in lines of antique simplicity that recalled the royal statues of Egyptian temples.

After we regained the track that would take us down the escarpment to Keekorok, we were able to talk again. I confessed that it had been a frightening experience because, at one point, I had thought the Masai was trying to hypnotize me. "It's not impossible," Juliet said, and she recalled something Joy Adamson had written in her portrait book *The Peoples of Kenya* about the witch doctors who sat for her and tried to mesmerize her while she painted them. Luke said that no matter where you hid yourself in Masai country, two or three natives were "sure to come up out of the ground"; he had never known it to fail. Lou examined the choker with admiring eyes and then tried it on my neck. It fitted perfectly and that seemed the most amazing aspect of our strange encounter.

I have the choker with me now and often I look at it with wonder. It has become a kind of talisman for me. It has the miraculous effect of reminding me of the great mysteries hidden in Africa's ancient, dark interior and of its hypnotic power to frighten and enchant . . . at one and the same time.

XI

THE ADORABLE HYENA
OF MARA MASAI

I T WAS EXCITING to find ourselves physically present *in* the Rift country after having gazed down on it from so many high places in Kenya. From above, one can see only the blinding shimmer of a seemingly lifeless land, but here, on the rolling grassland of the plain, we were among multitudes of assorted animals sheltering in the groves of acacia trees and thornbush thickets. Birds were in thousands . . .

Weary and drained from the ride down the escarpment, we made no effort to photograph the wildlife of the reserve on the way to Keekorok Lodge, where we were to stay for three days. And yet we were driving past giraffe, zebra, gazelles, warthogs and herds of leaping impala. And tempting above all, from his lookout atop an anthill, was a purple-flanked topi snorting at us — a "first" for Lou as well as me. And then another rarity — a spotted hyena lounging by its burrow.

Riding slowly past these marvels with our camera lenses capped, it occurred to me that we had become choosy after two weeks of wildly photographing everything that moved. One never wearies of looking, but one becomes selective about what to photograph. We had been out fifteen days by the time we reached Mara and already I had fifteen rolls, each of thirty-six exposures, packed in my suitcase — more than five hundred snapshots and still we had seen only a few of the larger mammals listed in our field guide. Impossible to photograph every wild creature you saw — and then there was also the *look* of the land to be recorded. . . .

The endless plain studded with flat-topped acacia trees rolled away to horizons of dim blue mountains. Infinitudes of space, boundless beyond measure, surrounded us on the way to Keekorok. Our eyes were strained by its endlessness. This was a land of incredible grandeur and it would go on and on, heaping grandeur upon grandeur until almost the end of our safari.

The Mara Masai Game Reserve in Kenya runs down to the Tanzania border, where it continues as the great Serengeti Plain with its famed national park covering an area of over fifty-six hundred square miles. Adding Mara's seven hundred square miles to this, you have a continuous wildlife reserve bigger than the combined areas of Connecticut and Rhode Island, bigger than the whole of Northern Ireland, and just about equal to the areas of all the islands in the State of Hawaii.

Keekorok Lodge, the only hotel in the Mara Masai Reserve, is set in a vast, shallow depression. Its large main building, rustic and luxurious, has a long terrace shaded by trellised Bougainvillaea. This faces a circular slope of lawn with surrounding cottages and a swimming pool. Our double cottage was the last in the semicircle at the foot of the lawns and on the way to it, we stepped around bands of gray monkeys nursing their babies on the stone steps leading down. Our cottage happily was supplied with every comfort needed to recharge our energies after our long exposure to the exhausting wonders of the open plains.

Two hours later, showered, powdered and dressed, we walked back up the lawn for a sundowner date with our escort on the café terrace. We felt, and hoped we looked, reborn. We had brought our cameras, in case there might be a chance at a shot of a superb sunset, known to be fabulous in this region of Africa.

We heard the lodge before we saw it. It was packed with Americans this night and sounded like a strident hen house. All the terrace tables were occupied, but in the lounge we found Luke in possession of a table with three chairs tipped against it, waiting for us.

I was astonished by the age of these American tourists. The majority were white-haired and could have been grandparents. I had thought that *we* were the only sexagenarians on safari in the whole of Africa, the only matriarchal old girls still tough enough to take the arduous life of the bush! But here were throngs of contemporaries, laughing and joking as they table-hopped about the crowded lounge, greeting sprightly comrades with glad cries — "You *made* it, by golly!" or "Didja see the lions today?"

The women all wore Masai beaded jewelry and flower print dresses with matching sweaters. The men were paunchy and balding. In gay cotton jackets and white permanent-press trousers, they had the well-heeled look of retired corporation chairmen. Their geriatric hilarity simply *had* to be forced (although it didn't sound so) and I kept asking myself, "How do they *do* these high jinks?" But soon I recalled safari advertisements I had read — "See All Africa in Twenty-one Days" — and, repentently, I began to feel a certain admiration for these group travelers who, contrary to our deluxe schedule that gave never less than two-night stopovers anywhere, were on safaris so fast-paced that they seldom had more than one night in one place. "Come hell or high water," in their words, they were out to see the wonders of their planet earth while still living on it and, more importantly, *while still mobile!*

Right here is the place for some general and personal observations of our fellow travelers in Africa. First, there is absolutely no age limit for safari travel. Anyone who can still move a limb can do it. If surface travel is too rough for aging bones, one can always fly from reserve to reserve and then make Land Rover sorties mornings and afternoons, when the wildlife comes forth to be viewed. There is also a most extraordinary medical service covering every corner of East Africa — the Flying Doctors' Society of Africa — which operates light ambulance planes to rescue members who fall ill or suffer casualties in any remote spot and transport them to one of Nairobi's excellent hospitals. It costs just fifteen shillings ($2.50) a month to belong to this protective society and the membership card — listing name, blood group and allergies — is a comforting ace in the hole when roaming the great African bush. . . .

Our elderly compatriots were already gone next morning when we were awakened for an early start on our first game run through the Mara Reserve. According to our parks guide, it was Mara that possessed the largest number of lions to be found in all Kenya. Out on the rolling plains, one could see from miles away where the lions were on their kills of the previous night, for it was there the Land Rovers and zebra-striped minibuses were parked in circles around the feeding animals. Luke scorned these crowds, saying, "We'll find our *own* lions!" And sure enough, in the course of the morning we found three different lion groups — the first was five males tearing up an immense buffalo carcass; the second a lioness leading her cubs through the thornbush; then the prize of the day, our third lion, discovered at noon. He was a huge male lying on his back asleep on top of an anthill, all four legs in the air, tail drooping down one slope of the hill and head dropped down the opposite slope, his pale, whiskered muzzle straight up and eyes tight shut. Luke drove around the anthill once, to make sure no lioness was lurking in the long grass. In a whisper he asked, "Which end do you want to start on — head or tail?" Choking back our laughter, we indicated the head.

The anthill was like a round-topped pedestal as high as the Land Rover, and when we stood up on our seats with our cameras to shoot from the open roof hatches, we were on a level with our mighty subject and within twenty feet of him. The noon sun brought out the titian tones of the mane that framed his face like a ruff. My lens showed a pink spot on the tip of his broad black nose and three flies on it that I longed to brush away, for his sake. When we drove slowly around to his posterior end, I gasped over the revealed wonders of this sleeping beauty. His paw pads were drooping in the air like wilted flowers and his neat testicles looked like Russian Easter eggs covered with lavender plush. Each time I clicked, I heard my excited inner voice crying, "You're *mine*. . . . all mine. . . ."

He woke on our second tour around him, as if he had gallantly responded to Juliet's moan, "Won't he *ever* make a move?" — for she was shooting in vain with her movie camera. He stretched, yawned, and suddenly sat up on his pedestal and became a Babylonian statue directly out of the courts of Assur-banipal, except for the missing wings.

Luke turned on the motor as the lion started down the anthill, coming toward us with his yellow eyes narrowed. I clicked him once before we swung away and caught him head-on in a kind of grumpy humor for having been disturbed in his siesta. But not disturbed by us, we immediately discovered, but by his lioness walking toward him from the other side of the mound. Her lean, honey-colored body was slowly moving through the yellow grass as we veered away.

Having spent more than an hour in a royal presence, we were exhausted as well as exhilarated, a most peculiar state to experience. Apparently one can get a real "high" on lions. I knew that I would recognize that particular lion if ever again I encountered him anywhere in the bush. I had taken him into myself with such visual and emotional intensity as to make him forever a familiar part of me.

This strange sense of possession is, as I interpret it, the essential value of "game viewing" and the reason why it can never pall. You come upon a wild creature never before encountered in your harassing modern life. You stay quietly with him as long as he allows you to. From whisker to tail you absorb his beauty and novelty

140

in a private state of rapture that resembles the devotion of an answered prayer. After you leave him, you discover that he is still with you — a permanent imprint, not necesarily of a lion, giraffe or elephant. He is a unique *personality* who has become part of you, and I might add, an enhancing addition.

As we drove away from our prize, I remembered some recent reading in the British paperbacks we had been buying in various hotel bookshops — memoirs of African game wardens, explorers and zoologists. In Richard Carrington's *Elephants*, I had come upon a passage that summed up the author's belief about the value of wildlife and the urgent need for its conservation, not simply for the self-interest of tourist revenue, but rather for a broader principle that involved the human observer of animals in the wild. Carrington wrote:

> The evolution of the human consciousness is gradually leading to a greater understanding of certain fundamental truths. One of these is that a compassionate interest in the welfare of animals less advanced in the evolutionary scale than ourselves is an essential component in the character of the truly civilized man . . . accepted by all thinking people as a simple and unassailable criterion of spiritual maturity.

This amplified my first formulation of what going to Africa really means. I saw it now as a search for some essential element in "spiritual maturity" that modern life, segregated from wild animals, cannot give. It also explained my mad involvement with Africa's animals. What the animals gave back to me, in return for my ecstatic delight in them, was to draw me into a new dimension of feeling and thinking, for which a vocabulary has not yet been invented.

Plumes of dust far out on the plains marked where other safari parties were searching for wonders. The "milk run" was engagingly convivial, I thought, and I wondered how many of those camera-hunters out there were discovering, as we had, the mysterious bond that unites mankind and the great mammals who, nearly two million years ago, evolved together in the vast cradleland that is Africa.

"We've got time before lunch," Luke said, "to make a run out to the hyena burrows. It's only about eleven miles from the lodge."

We had a tail wind part of the way and it blew our own dust plume back into the Land Rover. We covered our cameras with plastic bags and tied bandannas over our heads as the pale powder mercilessly drifted in upon us. Dust was a hazard of the dry-season safaris that we accepted without complaint, knowing that in October's "short rains" we could not move a mile without bogging down in the muddy tracks. Even when Luke maneuvered the car out of the tail wind, the dust still came up through the floorboards and invaded our nostrils. Our mouths we kept shut.

The hyenas turned out to be worth all the choking discomfort of getting to them. Their burrows were dispersed over a quarter-acre of stony ground, black holes with worn, rounded edges, out of which, here and there, protruded the head of a sleeping hyena pup. A few adult females lay about, watching the burrow entrances. As the car crept up to within camera range of a pup's muzzle sticking out of a dark hole, its mother got up and came to protect her young. She was a big spotted hyena with a very sloping back and a massive head set with large dark eyes and short, rounded ears. In her absolute ugliness and ungainly sloping lope, she had the feral beauty of *une jolie laide*. She paid no attention to our car hovering over her den with three cameras aimed directly into it. She snuffled her protruding pup, which probably was an order to quickly vanish downward. Then she turned around and backed her body more than halfway down into the hole, plugging up her litter in a maximum security unique and very touching to me. I felt the sting of tears when she dropped her head back and shut her eyes, feigning sleep in this gesture of absolute trust in us.

In my viewfinder her head filled the frame. The furry, closed eyelids bulged over the great orbs of her night-seeing eyes. Her massive jaws, the most powerful jaws of all living mammals, were blunt, black and closed, the dark lips curved upward as if she were smiling. I kept her in 200 mm magnification and shot continuously while waiting for something, I knew not what, to happen. It seemed extraordinary that the sounds of our three cameras — Juliet's movie whirring and Lou's and mine clicking — did not disturb her. I was watching the bulge of her eyelids with my finger on the release button when she suddenly opened just *one* eye and looked at us with a kind of comic coquetry that seemed to say, "Don't kid yourselves, I'm not asleep." For just two seconds that single dark eye gleamed at us, then closed. None of us spoke or even breathed. So she gave us one more pose that could only be called "adorable." She wriggled her body about until she lay flat on her stom-

143

"She suddenly opened just *one* eye . . ."

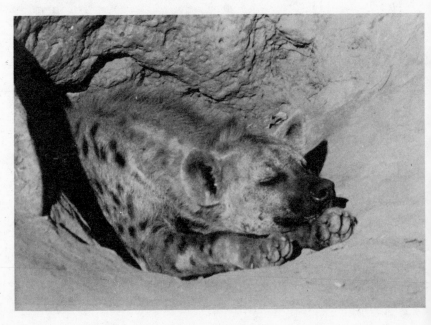

The spotted hyena flat on her stomach
with her body plugging the entrance to her burrow

ach, stretched her spotted front paws straight out before her, and laid her sleeping head upon them like a dog. . . .

Luke backed the Land Rover as quietly as possible away from the burrow, and our hyena did not open an eye to watch us go. At last we headed into the wind going back to the lodge and now there was no dust except the plume behind us.

Luke remarked casually, "The pups don't come out until after sunset. . . ." This left us hanging to the hope that he might bring us back to the burrows that same evening but unwilling to ask after his long day of hard driving. At some distance from them, he suddenly stopped the car and pointed down at a scattering of chalk-white forms, compact and capsule-shaped, like small breakfast sausages. "Hyena droppings," he said. "Almost pure calcium from the bones they crush and swallow." The droppings were the neatest and cleanest little objects imaginable and impossible to think of as excrement. Luke grinned at our astonishment and drove on, adding, "They always drop far from their burrows."

My admiration for the much-maligned hyena soared yet again. No writer I had read up to this point had ever had a kind word for the hyena. From the earliest days of the big-game hunters like Teddy Roosevelt, whose *African Game Trails* was published in 1910, the hyena was despised, feared, and usually shot on sight. In 1940, when Osa Johnson published her book about life in the African wilds (*I Married Adventure*), she had loving words for all the wild beasts of the bush, except the hyena. About him she wrote, "Both Martin and I developed what amounted to a positive loathing for the hyena

145

and shot him on sight, without feeling or regret. He is an ugly sneaking coward and apparently knows it, for he slinks along on his yellow belly . . . with never a show of spirit or clean, honest fight." Maberly's recent paperback guide, *Animals of East Africa*, reports that "natives sleeping unprotected in the bush at night have often had part of the face, or limb, bitten off by a Spotted Hyena." Even the primitive Africans associated hyenas with witches who were believed to ride about at night on their spotted backs and sometimes turn into hyenas themselves.

Riding away from the entrancing encounter with "our" hyena and talking excitedly of the pictures she had given us, I found it difficult to equate my own feelings about hyenas with the contemptuous reports of them I had read. If there was such a thing as hyena-love at first sight, then I was its victim or, more accurately, its recipient. Every time I thought of how that great ugly-beautiful beast had received us in absolute trust on the very threshold of her den, and entertained us like a worldly matriarch showing *her* way of protecting her young, I was thrilled with joy and gratitude.

It did not occur to me that our encounter had been exceptional in any way. Not until many days later in the Serengeti, when we came upon the only book that gives the spotted hyena his just due of understanding admiration, did I realize that few humans outside scientific circles have experienced the rare proximity we had with "our" spotted hyena at her den. In the Ndutu Lodge on the shores of Lake Legaja we found on sale Hugo and Jane van Lawick-Goodall's engrossing book *Innocent Killers*, written after four years of close obser-

A spotted hyena pup not yet with his spots,
like a dark teddy bear with a silvery face mask

vation, in their natural habitats, of hyenas, jackals and wild dogs. This study in depth is illustrated with Baron van Lawick's stunning photographs, most especially of the hyenas, which his camera reveals as *personalities*, playful, social, and exceedingly brave. Had I not already become a hyena-lover before I read it, their book would have converted me with its first page.

Just before sunset that same evening, our indulgent driver took us back to the burrows to see if any hyena pups were out. We had the hatches open and the sandbags ready. The sun had dropped below the horizon by the time we reached the den area and, in the pink afterglow, we saw more than twenty cubs sporting together outside their holes. Luke stopped at a discreet distance from them and gestured with his hands, "They're all yours."

The older pups already showed their spots, and the younger ones were still in the black coats they are born with, like dark teddy bears with silvery face masks. In the peculiar pink light on the barren ground around the holes, their shaggy bodies stood out clearly. Now we saw that the dens connected underground when a black pup emerged from one hole, wobbled over to another and disappeared down it to reemerge a few moments later from the first hole. Juliet, shooting with Eastman Kodak's new fast "after dark" film, was, she hoped, catching the constant movement. In and out, back and forth, they played, and when an older spotted pup (wearying of the game?) sat deliberately over an exit hole, the smaller body of a black pup could be seen squeezing out from under the spotted behind.

I set my camera solidly on a sandbag and shot with

my widest lens opening and the slowest speed I dared under the circumstances — one-thirtieth of a second, then one-fifteenth, praying each time I pressed the release. A hyena nursery doesn't come up twice in any photographer's life! Lou, shooting with fast black-and-white film, would have better luck, I was sure. But I was determined to try for the magical pastel colors of the scene — the pinkish den ground and the narrow foot trails raying out from it into the dim gold grasslands of the plain, where the parent hyenas were hunting the evening meal for their rollicking young. . . .

I shot until the spots on the older pups were no longer visible, wishing we could stay all night with them. Luke, using his headlights on the way to the lodge, now and again picked up the phosphorescence of hyena eyes shining in the dark, as they prepared for their night's hunt. I could only wish them luck.

XII

WE WERE
NOT "BIRDERS"

By the beginning of our third week out, the subjects each of us chose for our cameras showed our growing specializations — a form of self-preservation, since now on every side the bush offered more than we could take in without foundering.

Lou, for example, became a connoisseur of horned buffalo skulls lying about on the open plains (too tough even for a hyena to crush), of towering anthills with multiple spires and buttresses, and of elephant-wrecked trees uprooted beside the road. One day we came upon a dead elephant completely hollowed out by vultures and she carefully photographed this stunning evocation of slow death in the bush. The huge gray skin, bloated stiff by air and corruption, had kept the shape of the elephant-that-was, except for the tusks, taken by poachers or the game warden.

She also became a collector of game reserve signs — beginning with that old safari cliché, ELEPHANTS HAVE RIGHT OF WAY, which always startled and delighted us no matter how many times we encountered it. TAKE NO LIBERTIES WITH DANGEROUS ANIMALS. . . . STAY IN YOUR CAR

153

WHEN IN SIGHT OF ANY LIVING ANIMAL. . . . VISITORS ENTER
THIS RESERVE ENTIRELY AT THEIR OWN RISK. . . . STRICTLY FOR-
BIDDEN FOR PEOPLE TO WALK ANYWHERE OUTSIDE THE BOUND-
ARIES OF THE LODGE OR SAFARI CAMP, YOU ARE LIABLE TO A
FINE OF 1,000 SHILLINGS. . . .

Juliet, on the other hand, would ride for miles without using her movie camera. She was on the lookout only for the kind of wildlife she had not seen on her previous safaris — notably bongos, bat-eared foxes and lions in trees. Unfailingly she had kept intact the memory of more than a thousand feet of exposed film and she also continued to compose her travelogue as we went along, adding to it subjects of botanical interest when we came upon them. Trees in bloom always called for a stop, especially if they were the progenitors of our Hawaiian trees, such as the fiery *Erythrina* or the royal purple *Jacaranda*. These we found in Kenya, more vividly colored than their counterparts on Kauai but otherwise identical.

I had long since made my camera the surrogate of the writer's notebook. Each shot was an experience in total identification with my subject, to remind me later of the emotions I had experienced in its presence. The look of the land was now driving me wild with the never-changing changelessness, and I began using my polarizing filter to increase the contrasts of color in skies, grass-lands and distant hills. Because this filter was fairly new to me and tricky to use, everyone else was immobilized, waiting for me while I "created" my landscapes. But no-body complained.

As we traveled along, each pursued her special inter-ests and new vistas opened up for all. We marveled at

The author photographing a wayside wild flower
with an audience from a nearby village

the cathedrallike beauty of Lou's anthills and the mute witness of her bleached bones. We shared Juliet's excitement in her discovery of a small roadside hibiscus that was the plain little ancestor of all our ravishing hybridized Hawaiian hibiscus. As a triple mirror reflects three differing views of the same subject, we saw through each others' eyes an unsuspected marvel each time we stopped to take a photograph.

Only when we were in the presence of the great monarchic mammals did we react through our separate subjectivities, each eagerly seeking her own private bond with the lion, the giraffe or the adorable hyena.

However, singly and collectively, even without regret, we gave up on the birds of Africa about midway on our safari.

In the first place, we were not real "birders," as the safari escorts call the Audubon Society members who go to Africa mainly to do bird-watching. Save for birds en masse, as at Nakuru, or for the giants of the plains like the trotting Somali ostrich of the north, the kori bustard of the Serengeti, or the omnipresent vultures and their scavenger companions like the marabou stork, the bird life of East Africa would have been impossible for us ignorant ornithological amateurs to dare to tackle. It was an avifauna simply beyond our power to count and observe.

Early on, I had tried to count the species as they were named in the field guides of birds to look out for in the various game reserves. In Samburu Isiolo, the listed birds ran to more than three pages of fine print, with an average of a hundred and thirty-five different birds per page. In Mara Masai, I counted about six hundred listed birds and approximately that same number for the Serengeti — give or take a hawk eagle or a whydah or two. This advance counting was a self-discouraging thing to do. I began asking myself why one would *want* to name each bird in this "gorgeous galaxy of feathered gems" (as Roger Tory Peterson called it in his introduction to Williams's bird guide), or why one should struggle to learn what "birders" call the "field marks" of each. Was it not enough to simply look and enjoy, when they flew into our area of vision? Juliet and Lou felt as I did about the tremendous subject of Africa's birds, whose names in the field guide were the stuff of poetry: amethyst sun bird, tropical boubou, lyre-tailed honey guide, golden palm weaver, red-billed hornbill, fire-fronted bishop, superb starling. . . .

Luke was a "birder" and always drove with his binoculars on the dashboard shelf. Perhaps his long eyelashes owed their upward curl to the frequent pressure of his glasses against them. He would stop the car, peer through the glasses into a thicket, say, "Lilac-breasted roller," and then hang around waiting for it to fly. Once in Kenya's open woodlands we saw one. It alighted for an instant on the short grass — a spread of ultramarine blue wings, green back, lilac breast and long tail streamers. Until Luke named it, I thought I was staring at the biggest and most gorgeous butterfly God had ever created.

There was, however, a bird Luke despised. This was the five-foot-tall marabou stork. He called it "the undertaker" and could find nothing to admire in its heavy-billed head and naked flesh-pink neck, from which hung a long pale pouch. The marabou is a common bird in

East Africa, a scavenger that keeps company with vultures. I developed a fondness for this easily identified bird, whose sad appearance made me think not of an undertaker but of a broken-down tax collector in a worn brown overcoat. But when it flew, it was transformed, and only then revealed a true splendor with its amazing ten-foot wingspread. The bird guide said that as well as being a carrion eater, the marabou was also a welcome destroyer of locusts, which appetite, I suggested, made it a land conservationist worth honoring. Having become a champion of the under-bird, I now discovered hidden charms in the marabou, remembering its contribution to the Jazz Age, when its soft tail feathers were used by milliners to make long, slinky scarfs without which the Charleston could not be properly danced. "Bloody scavengers!" was Luke's reply to all my importunings in behalf of the marabou stork.

But he was to have his comeuppance with scavenger birds not long after our genial arguments over the questionable charms of the marabou. This event took place in Ngorongoro Crater, beside a shallow river we had chanced upon during a game run. The river flowed narrowly over stones except for a single pool at a bend. Here Luke pulled to a halt because he saw there was a peculiar bird activity going on, unknown in his experience.

From the sky, we saw scores of vultures dropping on their long, eaglelike wings to the opposite riverbank, folding their pinions as they landed. Hastily, they strutted down to the water and began to splash and duck their down-covered heads under the water. After fluffing their brown feathers in a thorough cleaning, they looked

158

like a gathering of shabby old men drying off in a public bath. Juliet all but wept for joy in the excitement of this extraordinary sight. "But absolutely *nobody* I've ever heard or read has ever reported that vultures bathe," she said. "Maybe we're the first ever to see such a sight." She turned to Luke. "Have *you* ever seen vultures bathing? Did you ever *hear* of such a thing?" Luke shook his head, unable to speak. The discovery that the "bloody scavengers" enjoyed cleanliness as much as he did seemed to have completely upset him.

Group after group of vultures flew in while we photographed the incredible scene. It was as if they were coming on rendezvous at that specific hour, as if summoned from the plains: "Calling all vultures! It is bath time!" Juliet panned her camera from sky to pool and back to the riverbank, where the cleansed vultures were drying themselves in the sun. "You must all remember the *time* of this happening, so we can tell all about it to the park warden," commanded Juliet. "It's fantastic. He may not believe us until he sees the film."

Luke was silent until I asked him how one would describe the harsh sound the vultures made when squabbling for a place in the pool. With maximum British reserve, he replied, "The bird guide calls it a squark. . . ." He let the ugly-sounding word sink in, then — lest we forget the more repulsive habits of our bathing beauties — he added, "You hear it loudest when they're fighting over carrion."

At Seronera Lodge in the Serengeti, bird-watching was the chief occupation of the outdoor cocktail terrace, where an outsize bird-feeder on a stone pedestal was kept heaped with bread scraps by the dining room boys.

From the bare trees around the terrace, elegant small birds dropped down in flocks, piling up on the feeder so thickly that many found standing room only on the backs of others. I listened to the "birders" around me identifying starlings, tits, barbets and shrikes and watched them make notes in their checklists on the When and Where of the sightings. In a way I envied them their single-minded pursuit, for Africa was beginning to wear me down with the impact of its mighty marvels, and the Serengeti nearly stupefied me with its immensity.

In Dr. Bernhard Grzimek's book, *Serengeti Shall Not Die*, published in 1959 by the German naturalist after he had counted by aerial survey the animal population of the Serengeti, he writes about the mystery of animal migrations:

> The plains of the Serengeti are said to harbor more than a million large animals, and these are constantly roaming in large herds. Sometimes there is one Wildebeest (Gnu) beside the other as far as eye can see; at others the same area is completely devoid of animals for months on end. There are many hypotheses about this migration. . . . Up to now, nobody has found how to follow the wandering animals. During the rainy season, one cannot drive even a station wagon over the few existing "roads."

The doctor and his son Michael taught themselves to fly a plane so they could make their Serengeti census from the air during the wet season, when the grass was up and herds on the move. They accomplished this heroic task for the Tanganyika National Parks, whose administration raised a memorial stone to Michael, killed in a

The Serengeti — a leopard on the lookout
high up in an acacia thorn tree

plane crash shortly after their census was finished. Bernhard the father carried on and wrote the book they had planned to write together to give the world its first warning of the danger of starvation of Serengeti's great wild herds if the Masai cattle were not restricted in the pasturelands they shared with wildebeest, zebra and gazelles.

The Serengeti that we saw at the end of the dry season was so totally different from what the Grzimeks had photographed in the rainy season that it might have been a scene from another planet. After our stop in Seronera in the middle of the Serengeti National Park, we headed south for Ndutu Lodge on the shore of Lake Legaja, where the van Lawick-Goodalls had made their study of the wild dogs Juliet hoped now to find and photograph in order to complete her safari film.

We were, of course, off the main track of rocky road that led from Seronera to Ngorongoro, and now were driving through what looked to be a gigantic dust bowl from which all life had departed, leaving only the memory of its multitudes in the footprints that patterned the dust. Though I knew that this was to be our last trek in the Serengeti, I felt it would never end. With horizons invisible in the shimmer of whirling alkali, the parched plains seemed to stretch beyond infinity. In the heat and dusty discomfort, we had long since given up trying to talk with each other except by gestures.

Now and again Luke pointed to a distant herd of wildebeest or zebra strung out across the plain, plodding single file with heads hanging low, following their leader, who seemed to know the way through this desiccated hellhole to water and grass beyond the hills. Then it seemed that we were seeing the Serengeti in its final death throes and those strings of beasts who had dallied behind the great main migrations appeared to be the last survivors of their species, passing in review before the eyes of man for the last time. And we ourselves who were watching might very well have been the last human beings left on earth, and ours the only safari vehicle kicking up dust on this plain as far as eye could see. Surely we were the only travelers crazy enough to make such a detour in search of wild dogs, I thought, and even if we did find a pack of this elusive predator, I, for one, would not be able to see them clearly. A peculiar, inexplicable affliction had caught me somewhere in this wilderness and caused me to see everything double.

It was easy to conceal my condition from my companions, all preoccupied with their own discomforts of heat, dust and the rough track. Though shown to be only some forty miles by the map, this last stretch of the Serengeti from Seronera down to Lake Legaja took more of my endurance than all the preceding mileage of our safari, which by then I calculated was nearly two thousand miles. The dust was so fine, it seemed to invade one's pores. The powdery desolation of the plains in drought was a spectacle of a certain type of death in nature, the scene pockmarked by the countless footprints of animals that had fled the region as if forewarned by a special intelligence mysteriously passed to them.

I had the peculiar sensation that some part of me was separating itself from my body. I became two entities, and the separated part seemed to be looking down on its other half with pity. I knew that I was sick in a way I had never been before. Whatever the strange "bug" I had breathed in with the Serengeti dust, all I could learn about it for certain was that it was an exceedingly fast worker.

I had hallucinations as I stared ahead through the whiteness. I heard Luke say to Juliet, "Ndutu's over there on the other side of the lake." I began to see trees ahead of us, reduced to small black sticks by distance. The sticks receded as we rattled toward them and I had to fight back my tears when I realized they were only a mirage. If I did not soon find a bed on which to lay my divided body, I knew that I was doomed.

The heat was white. The great plain was white. We were lost in a blaze of whiteness except for that ever-receding line of mirage-trees toward which Luke steered with maniacal fortitude, in the deluded belief that the lodge was behind them. There *was* no lodge, there *was* no lake. The big white hunter! The big white fool! In those last miles of utter misery, I hated everything about Africa as intensely as I had previously loved it. I saw it then as a cruel and killing land of blinding brilliance, a huge hoax of mirages that drew one onward with phantasmagoric promises of trees, a lake and a lodge that would never *never* materialize out of the churning burning white haze. . . .

The black sticks of mirage trees slowly became real trees, on a real lake of almost solid salt. Then a water tower high on spindly legs loomed in the haze, with Ndutu Lodge below. I staggered when I got out of the car. With Lou's help, I was just able to walk to the line of cement-block cabins. I was put to bed with a fever of 104.

It was strange to be a helpless wreck, alternately shivering and sweating, and coughing with a violence that threatened to tear me apart. Lou said it must be malaria, but I knew how malaria felt because I had once had a bout of it after Mexico. This was something I had caught from the Serengeti dust.

Luke brought me tablets, British antibiotics of some sort, to be taken at three-hour intervals. He said I'd be all right by morning. And I was . . . but still too weak to join the game run in search of wild dogs. For the first time in three weeks on safari, Lou and I did nothing but sit on our cottage porch, talking quietly together and simply looking at Africa. The shore of Lake Legaja, perhaps a thousand yards from the porch, glared whitely with its soda salts, a background for the dark bodies of zebra and antelope ambling toward a freshwater cistern the lodge manager had built for them. Staring at this engaging parade with my diminished sight, I realized the shocking swiftness of an illness striking in Africa, especially here in these dry plains filled with the menacing, penetrating and unnamed dust.

The emptiness around me I filled with the heroic figures of Alan Moorehead's *The White Nile*, which I had brought in paperback. This was the land the explorers of the 1850s had *walked* across, all the way inland from Indian Ocean ports to Lake Tanganyika and Victoria Nyanza, in search of the source of the Nile. Burton and Speke . . . Baker, Grant and Stanley . . . how often their diaries had reported sudden illnesses that required their having to be carried in hammocks slung from porters' shoulders. Without antibiotics in those days, they had to survive as they could through their often fatal mala-

165

dies. Otherwise they died of them as did Livingstone, among others.

By noon I was able to walk to the lodge and wait for Juliet and Luke. The lodge manager led us into the cool lounge, said his barman would soon be on duty, and assured us that our friends had surely found the wild dogs across the lake where he himself had seen the pack last night. Our host, the famous George Dove who had helped the van Lawick-Goodalls find the wild dogs they had studied for almost a year while camping close by, was an extraordinary man, tall, robust and silver-haired, with rosy cheeks and a moustache waxed to sharp points. He was a living example of those hardy British colonials who had left on this land the distinctive stamp of their life-style. He stood at the open side of the lounge facing toward the lake while we talked and I saw him in chiaroscuro against the whiteness of the salty background — a landmark of a man who seemed quite ten feet tall.

He said that our party was lucky, for had we arrived a week later, the dogs would have disappeared, not to be seen again until the next rainy season, when the larger migrations passed through this locality following the new grass and being followed by their ubiquitous predators. I longed to draw him out about his experiences with the writer Jane Goodall, but beyond calling her a "dear, gentle little thing," and telling us that they had some autographed copies of her *Innocent Killers* on sale at the bar, his British reserve kept him from speaking of his association with the gifted young woman to whom he must have seemed like a Great White Father. In her preface, Jane Goodall had thanked George Dove for his help in finding the wild dogs, repairing their cars, delivering fresh food, and sharing his precious water. She ended her tribute by calling him "a true friend, of the sort one finds only once or twice in a lifetime."

Like so many of the East Africa safari camps built in the time of British rule, Ndutu Lodge displayed trophy skulls and skins in its interior as well as its exterior decoration. Outside the lounge, hung on the trunk of a thorn tree, was an enormous bleached elephant skull complete with tusks, the hollow eye sockets turned toward the bar. A friend had brought Dove the skull from far away in the Serengeti. On the top of a low divider wall that separated bar from dining room stood another present from another friend — a Masai warrior this time. It was a dried skin mask of a lion's head, tall as a British guardsman's busby, with fur and shaggy red mane intact and two black eyeholes out of which the wearer could look when the stiff skin was set upon his head. I said if that happened to be the lion his Masai friend had killed with a spear to prove his manhood, as was the custom, then indeed it was a rare trophy.

"Perhaps so," said Dove, "but the greatest treasure in this lounge is this one small stone cemented into the divider wall." He led us over to show it. His friends, the Leakeys, had brought it from the digs at Olduvai, and that fossil footprint in the gray stone was more than three hundred thousand years old. His fingers hovered over the shallow hoof-shaped imprint as if it were too delicate to touch. "Louis Leakey said it might even be a half-million years old," said Dove. He twisted his moustaches as he strode back to his lookout place at the open end of the lounge. "You'll be passing Olduvai tomorrow on the way to Ngorongoro. Make sure your man takes you in to the digs. Rough road but short and worth every jolt of it."

Olduvai! The renowned name was not written into our itinerary and I had not realized it was only a short detour off the road to Ngorongoro. This announcement quickened my heart like a shot of adrenaline. I had known about Olduvai ever since 1959, when Louis Leakey and his wife had uncovered the remains of the oldest manlike creatures yet discovered on this planet earth — a skull definitely not of the ape family, but of a hominid of the family of man, estimated to have lived at Olduvai nearly two million years ago. Africa, rather than Asia, was thus proved to be the cradle of the human race.

Olduvai . . . the "place of man's arising," I thought, remembering the most moving phrase of Gurdjieff, my teacher and guru of long ago. *Zinjanthropus*, the Leakeys had named their find, "Zinj" for short. He was a "near man" and after him they found the successor who became true man out here in a world of giant animals. I stared enraptured at the fossil footprint of the unknown beast that had walked this earth with my ancestral cousins of a half a million years ago, now extinct like "Zinj" himself.

Olduvai . . . over there to the east beyond the lake. I saw a white plume moving along the white shore across the salt lake. "There's the rest of your party coming in," said George Dove. They came directly to the lounge and Juliet's shining face told the story of their successful morning's game run. They had found a pack of pups at play on the plain, about a dozen in all. While they were taking movies, three adult wild dogs returned from their hunt and regurgitated the chewed meat, which the ravenous pups gobbled up almost before it touched the ground.

XIII

"WHO WERE THESE MEN?"

OLDUVAI, THREE AND ONE-HALF MILES off the main Serengeti-Ngorongoro road, is the drabbest-looking place in all of Tanzania, but the most thrilling spot to be in, providing one can realize its emptiness and imagine what the first men who emerged from this wasteland must have been like. Thanks to the inspired labors of the Leakey family — Dr. Louis, his wife Mary and their three sons — the long road back to those prehistoric men-creatures, who lived here nearly two million years ago, has already been mapped for our imagination.

Called "the Grand Canyon of Evolution," Olduvai is in fact only a river gorge hardly a mile wide and no more than three hundred feet deep. It opens up so precipitately on the grassy plain that one could easily walk over the rim and fall into it if not aware of one's steps. This was almost how Olduvai was discovered in 1911, when a German entomologist, Professor Kattwinkel, chasing a butterfly over the plain, nearly fell into the gorge, fortunately not all the way, but far enough to observe, and later report on, the rich fossil deposits in the layers of the walls.

We came with dramatic suddenness upon the Olduvai rim and Luke pulled up just in time. When the dust settled, we saw a guide's hut, a small museum and a ticket office. A sign forbade any but the park jeep with an official guide to descend into the gorge. However, Luke said he might "talk the chaps" into letting him drive his own Land Rover down to the digs.

While he negotiated, we walked out to a thatched pavilion behind the museum for our first view of the floor of the gorge. It was an irregular-shaped depression enclosed in places by walls or outcrops striped horizontally by bands of sediment laid bare by erosion. These bands were closely packed, of many colors and textures, made from volcanic ash forced through clay to sand. This "layer-cake" effect of the sediment stripes in the enclosing walls is the wonder of Olduvai, quite impossible to realize no matter how long you look at it. Every seventy feet of those layered deposits represents a retreat in time of five hundred thousand years, beginning from the top. When you are down two to three hundred feet, on the gorge floor, you are where the remains of *Zinjanthropus*, the "near man" and *Homo habilis*, the true man, were found, with the first man-made tools. You are now back in time about one and three-quarter million years. We keep telling this to our dazzled brains, which are unable to comprehend it.

The museum's tin roof covers curious photographs, charts and scale models of excavation sites for understanding these incredible comparative statistics. The bone bits found at the different levels show what creatures accompanied man in his slow, innovative evolution into a species now separated from the apes. Fish bones in this level, a bit of crocodile skull in this one, the foot bone of a manlike juvenile here and five teeth from the jaw of a prehistoric child there . . . you gaze in wonder at these fragile finds and at photographs of the finders flat on their stomachs under parasols, or on their knees inching (as Louis Leakey reported) "over ground as rough as a nutmeg grater" in search of the remains of the rodents and birds that their "near man" caught by hand before he invented hunting tools. One does not need to be an archaeologist to be overwhelmed by these finds.

Postcards for sale at the museum include a colored photo of a restored skull of "Zinj" and a black and white one of Dr. and Mary Leakey sitting at a camp table covered with fragments of the skull, which they are piecing together as calmly as if working on a jigsaw puzzle. These photos are talismanic in effect: they transport you directly to the miracle of the discovery of these prehistoric bits and pieces. In his *National Geographic* article (October 1961) reporting the find, Louis Leakey stated that their chances of uncovering even one bit of human bone in this ancient gorge where antediluvian carnivores once roamed were about one in a million. The discovery, as I see it, is half of the miracle. The other half was the Leakeys' intuitive certainty that they were on the right track of prehistoric man there in Olduvai. The first time they came upon the gorge was in the early thirties and their certainty is surely dated from the moment when Louis Leakey picked up a rough stone flaked to fit the human hand and exclaimed, "Who were these men?" as he examined the implement to determine if it was fashioned for chopping, cutting, or scraping.

Luke called us back to the Land Rover as soon as he brought our official guide. He was a young African in a white business shirt who gave us a short lecture on the layout of the digging sites before we started down. His scientific vocabulary was loaded with anthropological, geological and paleozoological terms, which he pronounced slowly and meticulously as if waiting for us to keep up with him.

"We call the different levels *beds*," he said in conclusion. "The bottom one, where we are going, we call Bed I. That's where we found many of the Lower Pleistocene fossils."

The track down to the floor was steep and rutted. Dust billowed up from every pothole we jounced through. I saw a sign requesting visitors not to pick up any souvenirs around the excavation sites and when we were on the valley floor, I saw the *raison d'être* of such a sign. Small rocky fragments were scattered about temptingly, stone indistinguishable from bone except to the trained eye. These fragmentary bits were bleached gray-white not only by age but also by the volcanic ash of the strata from which they had been excavated.

Chalky roads spread out in all directions over the valley floor, not flat, as it had looked from above, but humpy, its low hills spotted with thorn-scrub and cactus. Beyond the hills one caught glimpses of the walls — those fossil-bearing exposures that stretched on for a hundred and fifty miles in the main gorge alone.

Our guide directed Luke through the web of roads that connected the various digging sites and brought us at last to the one that had made world history in 1959 — the place where "Zinj" had been found, surrounded by his tools. Luke parked off to the side and we reverently approached the scene, appropriately on foot. The site was no bigger than the stage of an average theater, and dramatic in a natural way. The backdrop was a dark "layer-cake" wall with its geological strata clearly banded and in many cases picked clean of its incredibly ancient, imbedded fossils. The stage was flat, shaped like an apron, and in the front stood the single prop made by man in this archaeological amphitheater — a cement pedestal holding a stone plaque engraved with the words:

THE SKULL OF

Australopithecus boisei

(*Zinjanthropus*)

WAS FOUND HERE BY

M. D. LEAKEY

July 17th, 1959

Three stone hand implements were placed in front of the sign, examples of the artifacts found with "Zinj's" bones.

Our guide gave his speech as we stood about in the dazzling morning sun, which would soon make a hell-hole of the gorge. How had the Leakeys been able to endure all those years of intensive labor in such a place, most often on their knees? They worked necessarily with dental picks and paintbrushes to uncover the delicate detritus of prehistoric man, picking and brushing

177

and picking again unendingly. With what intelligence, what genius, the Leakeys had postulated, from objects found with his bones on his cave floors, the daily life of their earliest man! Inspired guesswork? Enormous scientific knowledge? Probably a combination of both. I remembered that the doctor and his wife were not only great archaeologists and prehistorians, but also that both were the possessors of an uncanny intuition. I had read everything published about their labors in this desolate gorge, and, remembering, turned away from our guide so that I could listen to the silence of this unique place and have my own new and wonderful thoughts and, of course, emotions.

The distant layers of the walls were unbearably moving. They wore the story of man's arising pressed into their strata like words on a page, words composed of bone and stone and fossil imprints. Millions upon millions of them lay here, still uncovered, waiting to be read. The creatures that had accompanied man in his evolving centuries were all believed to be there. Sometimes the Leakeys interpreted "random bones" discovered far from their working site or brought to them by their native assistants, and by comparative study determined these to be skeletal parts of vanished monsters resembling elephants or horned rams, much greater in size than any animal known on earth today. Man had emerged into a natural world of natural giants.

It is here in Olduvai, on the flank of the Great Rift Valley, that Africa takes you deepest into her secret, dark and mysterious origin. In this gorge she has split herself open like a sectioned geologic chart, and thereby reveals her millions of years of continental existence and the kinds of lives she created in her experiments of each epoch. To me, her most poignant revelations were placed in the strata where she gives her evidence that man and mastodon evolved together, for better or for worse, willy-nilly.

At the base of the wall that backdropped the excavation site were some small piles of dirt that had been sieved through a fine screen to retrieve from "Zinj's" eating floor minute bone fragments, so tiny and delicate that five or six could fit under Leakey's thumbnail. The Leakeys had left the site as it doubtless was when we saw it — scraped, brushed, and sieved up, every last bit of corroborating evidence gathered by them that could prove their prehistoric man lived right here at this precise place.

Our guide, concluding, raised his arm and pointed up to the level where Mary Leakey had found two human teeth at the end of a day of frustration. "The first fragments of the skull of *Australopithecus boisei*," he said. "The date of the find is carved on that stone sign." I snapped his picture, his arm outstretched in solemn salute to a million-year-old ancestor. He had come a long way from the natives of Louis Leakey's early days at Olduvai, those first Kikuyus who refused to believe that the stone implements their master unearthed could have been used by prehistoric man. "Who could cut up a kill with a stone?" they asked. Dr. Leakey thereupon gave them a demonstration.

The doctor believed that he must master the implements of the prehistoric toolmakers if he was to understand Stone Age man, and had long since taught himself to make and use stone tools. Now he showed his assistants by laying out a ram's carcass, striking two stones together, and producing a flint-edged chopper with which he skinned and disjointed the ram almost as quickly as his cook could have done it with knife, cleaver and saw. The *National Geographic*, covering Olduvai at the time, photographed Leakey's exploit, which was accomplished, they stated, in less than twenty minutes! In the photograph, Leakey looks like a big white-haired boy in coveralls showing off before a group of Africans squatting on their heels, their eyes on his stone knife, black faces mesmerized in wonderment at the efficiency of its action. But there is much more than a seeming show-off prowess to be read into that documentary picture. This was how Leakey created *belief* in the natives he wanted to train as assistants — belief first in the use of stones as cutting tools, and then, inevitably, in the prehistoric men who had chipped them.

That Leakey was rewarded in his efforts has already become archaeological history. In the early sixties, Heslon Mukiri, an African who had worked for more than thirty years with him, discovered the first fossil fragment of a new member of the even older group that included both apes and man — *Kenyapithecus wickeri*, named for the owner of the Fort Ternan farm in Kenya where it was discovered. This jawbone fragment of one of Nature's failed attempts to produce man was eventually judged to be some fourteen million years old, which made *Zinjanthropus* seem a comparative newcomer on the evolving human scene.

We drove to other excavation sites that had been prepared for the visit of archaeological amateurs like us, but none to compare in interest to the "Zinj" stage, even though each added some clue to Stone Age man's history. On the way back up the track to the rim, I asked our guide how he had found a chance for such an extraordinary education in archaeology. "In the museum in Nairobi," he said proudly.

I recalled the unfinished display in the new Olduvai Wing of the Coryndon Museum. Our guide must have been one of the young men working there under Leakey's instruction, setting up the remarkably realistic dioramas that told the story of prehistoric man. I longed to ask him many questions about his work with such a famous archaeologist, but not in a Land Rover lunging up a steep track full of potholes. Moreover, I was sure that he would have refused to give me any personal data on the Leakeys. When we had first met him that morning, I had asked eagerly if the Leakeys were working there that day, he had said no, they were away, he didn't know where, and had turned from me abruptly to show the subject was closed. The museum guides had been equally evasive. All Leakey's associates seemed to be in agreement to keep a protective screen around him, against us time-wasting tourists.

Although I secretly admired their protection of their hardworking chief, I still wished for just one glimpse of the Leakeys hunting fossils down there on the stony old floor of Olduvai Gorge where, some twelve years earlier, all the world's anthropology textbooks were suddenly outdated because of one small bone no bigger than a tooth.

As I was writing this chapter, I heard the announcement of Dr. Leakey's death on the radio. The indomitable searcher whose excavations had added more than a million years to man's known presence on this planet died on October 2, 1972, in London of a heart attack at the age of sixty-nine. This news so saddened me that I thought of deleting the Olduvai chapter in its entirety. After all, I had never met the great man or even seen him from afar. All that I had to write about was a stupendous stage set, lacking the star — a stage set now become a memorial.

A memorial. That one word decided me, and I went on to finish writing about Olduvai as if Louis Leakey were still there, as indeed he was even when I was, though unseen. Olduvai was already at that time his memorial, every smallest scattering of stones swept aside on an excavation site was witness of his painstaking genius. He is everywhere in the gorge. Generations of archaeologists to come will doubtless be sifting those fossil-bearing strata said to be among the richest fossil grounds on earth; but no matter what undreamed-of "trial man" may yet be brought to light, the gorge of Olduvai will remain the stupendous memorial to the man of our time who first picked up a flaked stone there and said, *"Who were these men?"*

Olduvai vanished behind us as suddenly as it had first appeared. At twenty yards away from the rim, it had ceased to exist. The great plain surrounding us was an unbroken stretch of dry grass lifting gently toward Lemagrut in the south — a ten-thousand-foot volcanic peak seemingly composed merely of violet blue atmosphere. Beyond it, and about twenty-five miles distant from Olduvai, we would ultimately come out on the rim of Ngorongoro Crater, where we were booked for the next two nights in Crater Lodge.

"We'll just make it in time for drinks before lunch," said Luke hopefully as we started the long, slow climb up into the stark grandeur of the crater highlands.

XIV

NGORONGORO—
A WORLD APART

N GORONGORO IS THE CLIMAX of one's African experience no matter where planned in the safari itinerary — whether at the start, the middle or the end of it. This is the one place that every tourist, regardless of sex, age or special interests, praises ecstatically and hopes to return to "the next time around." Why does it seem so rare, as if of another planet? Perhaps because from no matter which direction you approach Ngorongoro, you must climb *up* to a misty altitude of seven thousand feet in Tanzania's crater highlands. Being so high and so cool, it seems to belong to another continent far from equatorial Africa.

I have heard travelers assert that Ngorongoro was the true Old Testament site of the Garden of Eden and that one day documents to prove this claim will be found, just as the Dead Sea Scrolls were come upon at Qumran. Ecologists say that Ngorongoro, with its year-round grass and water, enough to support a population of over ten thousand large herbivores and carnivores, is the most ideally balanced ecosystem of a prey-and-predator community to be found on earth.

The crater floor, Ngorongoro

Be that as it may, I will state here without hesitation that I believe every claim made about Ngorongoro, including my own fixed idea that it was the inspiration for the American painter, Edward Hicks, when he was producing his series of Rousseau-esque paintings, all entitled *The Peaceable Kingdom*.

The spectacular crater of Ngorongoro is a caldera, a circular depression twelve miles wide and surrounded by two-thousand-foot mountain walls, the remains of the collapsed volcanic cone. The floor covers about one hundred square miles — a mere patch on the game-reserve landscape compared to Serengeti's six thousand square miles, but a patch of absolute perfection. Because you cannot see anything of the crater until you are on its rim in an open, treeless place, you feel a shock of awe at your first sight; and, like the view you would have from an airplane far below into the bowl of dazzling atmosphere, nothing from so far above is seen clearly — only a pale wash of blue that might be a lake at the bottom.

The oldest game lodge in East Africa — Ngorongoro Crater Lodge — is built so close to the edge of the rim that it has nothing in front of it except the crater. This you can see almost in its entirety from the west windows in the cocktail lounge and restaurant, from all parts of its grounds — which are furnished with stone benches facing the view — and from the porches of the detached cabins built in a semicircle on the hillside. This arrangement is ideal for a great natural wonder like Ngorongoro Crater. Before descending, the visitor gazes down enraptured from above, as into a gigantic cauldron filled with marvels. Thus he becomes accustomed to such spaciousness before experiencing it at closer range.

We looked at the crater unwaveringly all our first afternoon. Meantime, Luke went off to service the Land Rover in preparation for the next morning's descent, which was to be an all-day trip. Beside us while we gazed at the volcanic masterpiece were booklets about it bought in the company store. These little park books, liberally illustrated with maps and photographs and written by specialists, are of immense value to the harried tourist. They give a brief introduction to what he is about to see and also help him afterwards to keep the reserves separated in his mind. Ngorongoro was covered by three separate booklets — "Geological History," "Animal Life" and "Ngorongoro's First Visitor," which was just what I was wishing for — the story of the first man from outside Africa to have seen the crater.

Dr. Oskar Baumann, an Austrian who was exploring East Africa in 1892 for the German East African Society, was Ngorongoro's first visitor. His diary was published in Berlin in 1894 under the title *Through Masailand to the Source of the Nile*, from which excerpts were translated for the text of the engaging little "First Visitor" booklet.

Baumann's foot safari approached Ngorongoro from the east, after climbing the Rift Wall at Manyara. With warrior Masai for askaris, they followed the ancient cattle tracks of this pastoral people (still inhabiting in restricted numbers the grasslands of Ngorongoro today) and on March 18, 1892, after a long push through mountain woods, Baumann made this entry in his journal:

> At noon we suddenly found ourselves on the rim of a sheer cliff and looked down into the oblong bowl of Ngorongoro, the remains of an old crater. Its bottom was grassland, alive with a great number of game; the western part was occupied by a small lake.

Baumann led his party down the steep slope and pitched camp at the foot of the precipice, moving on next day to a better camp made near "a small wood in the shade of a giant tree," and now his cool German prose warms as he describes what a "paradise" he had found himself in, "with terrific herds of wild game roaming in the wide plains . . . hardly shy at all . . ."

This was exactly what we were going to find down in the crater — a lack of shyness in the wildlife there, an innocent trust that the animals manifested in their long, thoughtful stares at us, fearless and faintly curious — as if no animal had ever suffered from the hands of humans! As if centuries of man's rapacious hunting solely for hides and horns, tusks and tails, had been wiped clean from the animal memory! As we sat on the rim reading about Baumann's adventures in the crater, we did not know that we were going to see — only slightly changed by a diminished wildlife population — the Ngorongoro that he saw seventy-nine years earlier. Such is the enduring quality of crater life.

Late that afternoon, however, we had a foretaste of the lack of shyness in the Ngorongoro animals that Baumann so often spoke of in his journal. We were loafing on our cabin porch, waiting to see how a sunset would transfigure the crater, when suddenly three zebras came over the rim through some shrubbery and began to graze a few yards from us. Our excited whispering did not disturb them, nor my gestures as I reached for my camera.

Ever since I had seen my first zebra running wild in Africa, I had longed to get close enough for a shot of one from the rear. There, where his stripes flowed together down the ridge of his spine and then seemed to be *braided* into the skin that covered his neat, bony tail, lay his visual charms. But I had longed in vain for such a close-up. Zebras encountered on game runs elsewhere had been too quick and nervous for our car to get near them. Moreover, they seldom turned their backs to a source of disturbance but stared at it fixedly, until on some silent signal the whole herd would swing about and gallop off — a stampede of striped backsides swirling in gorgeous, psychedelic designs impossible to photograph.

Now, here I had three of these enchanting horse creatures making themselves at home in my front yard, so to speak, the nearest I had ever been to *Equus burchelli*. They showed no alarm as I stalked them to within ten feet, eye fixed to my viewfinder, now filled with plump, striped rumps. Juliet said in a low voice, "Don't get too close . . . they're stallions . . . they kick and bite. . . ." At this, they raised their heads all together, hairy ears stiffly erect and black muzzles quivering. They stared at me — their great black eyes set off by the sweeping curves of their facial stripes, no two patterns alike. By then the good light was gone, but I held the zebras in my glistening viewfinder until they faded away among the trees, cropping the grass industriously as they went. They were headed in the direction of the crater. Would one of them become the dinner of a pride of lions that night? Or of a hyena pack? A haze of grass-fire smoke now covered the crater, like a curtain drawn over the scene of the carnivores' nightly hunt.

The next morning our crater was filled with fog, which Luke said would burn off by nine o'clock, giving us plenty of time to prepare for our day below. The air of the highlands was sharp and chilly. The Masai servants of the lodge, trotting around on housekeeping duties, wore European jackets over their cotton shirts and shorts, and their bare stork legs were blue-black with the cold. Juliet, who always included a weather thermometer in her safari essentials, told us the temperature was 52 degrees Fahrenheit. After breakfast we waited for Luke in the main lodge, where a log fire was crackling in the wide stone fireplace. At last he came and we started on our most revealing and exciting adventure.

Two one-way tracks connect the rim with the crater floor, one for the descent, one for the ascent. They are real mountain switchbacks, hacked in zigzag down the crater wall. On the way to the gap in the rim forest, where the descent track began, Luke stopped before the memorial stone that marked Michael Grzimek's grave on the spot where his plane had crashed on completion of the animal census he and his father made of Serengeti and Ngorongoro. It was a small stone with a few dried flowers at its base. The inscription read:

MICHAEL GRZIMEK
12/4/1934 – 10/1/1959
*He gave all he possessed for the wild
animals of Africa, including his life*

The grave, lying just below the rim road, with a hazy vista of crater behind it, penetrates one's heart with sadness for his fate as you inevitably think: "Another few yards higher and he'd have made it."

Slowly the crater revealed itself as we started down the zigzag track through layered mists, so bright with the sun behind them. First we caught glimpses of it through holes in the fog, like reflections from the bottom of a well. The Land Rover bucked over big stones on the loose dirt track and we hung onto guardrails with both hands, craning our necks toward the abyss of brightness that occasionally opened to show a grove of miniature thorn trees below or a string of unidentifiable animals no bigger than beetles from our height above them.

Then, quite unexpectedly after hardly a half hour of our plunging, angular descent, we were on the floor, riding over a golden grassland beneath a blazing sun. The animals that had looked from above like beetles turned out to be the large antelope called wildebeest or gnu. A line of them with horned heads lowered, with their long white beards and black shoulder manes blowing before them, leisurely passed us, reminding me of an ancient temple frieze in slow motion. We had never managed to get near to the wildebeest in the Serengeti, but here in Ngorongoro they not only paraded for us but, when they stopped to graze, they even posed now and again, lifting their bristly muzzles to stare at us while they chewed, "hardly shy at all," as Baumann had said.

The realization of being in a world apart increased the farther we roamed over the crater floor. The encircling walls take on a delicate shade of blue as you come toward the middle of this natural preserve. The volcanoes of the crater highlands loom up behind these walls and cause them to look much lower than two thousand feet — no more, in fact, than a circle broadly blue-penciled around the horizon, and this magic circle encloses all living things there — birds, beasts and humans — in a place of trustful peace that is palpable, like velvet to the touch. No animal lowered its horns or reared up in alarm at our approach. All stood their ground and examined us in a trance of reflective curiosity, as if taking the measure of a familiar species known to be harmless. Afterward, they went on with their grazing or bone-gnawing as if we weren't there at all. Even the skittish and prancing gazelles — Thompson's, Grant's and impala — gave us that motionless moment of intensive scrutiny, their dark eyes set like polished gems in white circles or lateral facial stripes.

Luke drove with angelic legato, easing us up to herds of wildebeest and zebra, then letting his motor die

A troop of zebra drinking

Rhino cow and calf, Ngorongoro

"Long, slow stares so deeply affecting"

Honeymooning lions

soundlessly while we stared back at the staring animals. We were so moved by their seeming trust in us, we were unable to speak and reluctant to break the spell by aiming a camera at them. Somewhere in the back of my mind lay a plausible theory of why the animals looked at us with such peculiar intensity. It was a formulation from long ago and far away. I made no effort to reach for it, knowing it would drift to the surface in its own time.

A troop of zebra drinking from a stream gave us close-ups of their decorative backsides after a long, speculative inspection face to face. A lioness studied us with her yellow eyes, then called her five cubs out of the riverine shrubbery and led them single file across our track out to the open plain. A few magic miles later, we came upon a two-thousand-pound rhinoceros with a calf at her side, both browsing on a low shrub. The huge female let us come up to within some thirty feet, then she swung about and faced us, standing stock-still on her stumpy legs, her nose horn projecting forward and up from between her small eyes buried in baggy folds.

It was at this exact moment that the formulation suggesting why animals look at us with such intensity floated up to the surface of my mind. I had heard it twenty-five years ago in Paris, in the days when I counted myself one of the elect permitted to sit before Gurdjieff as a disciple of his teaching. His words were clear in my memory; I heard his heavy, rumbling voice saying, *"The animals are waiting for us to move up so they can follow."*

I had my camera to my eyes when I heard those words again, a statement now out of context, never completely understood at the time, yet retained nevertheless for this moment when it would be. The staring female seemed to be taking me in through smell and hearing as well as by sight, reputedly poor in the rhinos. I snapped her head, that gray eminence that had survived unchanged through countless millennia of planetary life before Man arose to take his place in the evolutionary octave, a place above all the life-forms that had preceded him and occasionally, as with the apes, had prefigured him, a place beneath the unimaginable levels of Conscious Being that lay above him. *The animals are waiting for us to move up. . . .*

This was the answer I had been unconsciously groping for ever since my first confrontation with Africa's wildlife. This surely was why the animals' long, slow stares had been so deeply affecting to me even from the beginning of the safari. With their great, bifocal eyes they took us in, unaware that they were waiting for us to "move up" that ladder that Jacob saw in his dream, thronged with angels moving up and down, foretelling the immutable laws of evolution and involution.

I felt a thrust of compassion for our rhinos, cow and calf, who had permitted us to stay close enough to hear the snap of stems they browsed and to see the action of their triangular muzzles and prehensile upper lips that grasped and pulled. How many hundreds of pounds of vegetation must they pull up daily to sustain their enormous bulks? Locked into their place in the chain of life (because there was no room above? because their specific evolution had come to dead end?) they were held in servitude to their monstrous bodies.

I knew I was thinking anthropomorphically, a cardinal sin in biologists' eyes, a "sin" I rejected totally. How could one *not* be anthropomorphic in the presence of Africa's animals — ascribing to them *some* human attributes and feelings, when one realized oneself to be linked into the same life-chain with them? The place of the linkage made no difference; all one could know with certainty was that it was somewhere above, say, the rhinos.

I put aside my camera and began to study the gross, unlovely form of a beast that had emerged unprotected out of swamps and, as those swamps had dried up, had learned, through unimaginable millennia of slow adaptation, to endure on dry land. I watched a little red-billed oxpecker climbing up the cow's flank, nipping at ticks and blood-sucking flies, and the sight made me unaccountably glad — that is, until I analyzed my emotion and found that I was totally identified with the colossal discomfort of being a rhino cow!

She could have attacked us instead of allowing us to stay within sound of the snapping stems she browsed. The mammal field guide warned that the black rhinoceros (*Diceros bicornis*) is ill-tempered and sometimes charges without reason, especially in areas where it has been disturbed; in that case it is vicious and should never be taken on trust. But the rhino had always been "disturbed," I recalled. Wherever man had come into its domain, the rhino had been hunted and poached (even occasionally in the protected reserves) for the sake of its horn alone. Rhino horn was highly rated in the Orient as an aphrodisiac; ground into powder, it sold for as high as fifteen dollars per pound (five times the price of ivory) to the illicit traders around the wharves of Mombasa.

Our rhino's front horn hooked out at least two feet away from her, a singular defense for a false horn. Actually, it was a horn-shaped dense mass of hairy filaments compactly growing up from a bony base to form this version of a nose. Yet, with that strange weapon of compacted hair, she can gore a lion or dent the steel chassis of a Land Rover. She had virtually no active enemy on earth except man.

I wondered why she had caused me to recall the words of Gurdjieff. Was it her sheer *preposterousness?* Why had I not remembered that wise formulation earlier? When my first giraffe had looked down on me with its long-lashed eyes of curiosity? Or when the spotted hyena with head on paws had looked up and captivated me with her eyes of trust? Or, above all, the many times I had looked a lion in the eye and felt, always, an electric shock of anticipation for some nameless awareness about to be revealed?

"Why now, so belatedly?" I asked her silently.

She looked up, but not at me. A Land Rover loaded with vocalizing Italians had driven up and parked beside us. Their unsuppressed exclamations startled all of us from our private reveries and the rhinos from their peaceful browsing. With puffing snorts, they swung abruptly and scattered away at a bouncing trot, their ridiculous, short-tufted tails like corkscrews straight up in the air.

Watching them go, I knew that another dimension had been added to my African experience and that the direction of that extension was inward. . . .

As the sun climbed toward high noon, we and all the larger mammals in the crater sought the shade of the occasional thorn-tree grove. Luke drove us into some foothills beneath the rim to one of the authorized places where visitors could get out of their cars to stretch their legs and lunch on the ground. Three other safari cars were parked under high trees filled with birds, which the people, obviously "birders," were feeding and photographing as they swooped to snatch at sandwiches.

Luke chose a spot under a vacant shade tree at some distance from the "birders" and set up our party in the privacy to which he had accustomed us. He nosed the Land Rover in so that its flat front hood could serve as beverage bar. After standing all morning on the car seat to photograph from the open roof hatch, Lou and I sat down for our lunch while Juliet, who had sat all morning, ate hers standing up.

Our picnic place was at an elevation that enabled us to see over a large area of the crater's grassland, where wildebeest ambled in long, wavering black lines. Off to one side, between us and the floor, were some granite outcrops topped by stunted shrubbery, which gave good cover to the lions, hyenas and jackals whose predatory eyes, I was sure, were watching those herds as intently as I was, but for a quite different reason. Those other eyes were reading the menu for the evening meal. When darkness fell over the crater, it would become a place of terror and death for those large, gentle grass-eaters — wildebeest, zebra and gazelles — that stayed together all night in herds out on the open plain for mutual protection. Although I had heard that the predation sorted out for the most part the sick, the maimed and the aged, thereby "improving" the breed as was said, I had no wish ever to see a killing.

It was enough to hear the shrieks, snorts and grunts of the dying in the awesome threnody of the African night . . . it was enough to know that the lion gives a killing bite at the back of the neck or suffocates its prey by clamping jaws around the muzzle . . . that hyenas and wild dogs kill by disemboweling . . . and leopards by gashing open the throat. . . .

Even as I thought of these shocking, mercifully swift acts, I saw them now from my revelation, my new point of reference. They were but the efficient techniques the animals themselves had devised in order to keep their species alive on the great life-chain, alive and in their place while *waiting for us to move up!*

With that sweeping statement now fixed in my mind's depths, the impact of Africa on all kinds of people — from safari novices like myself to experienced old Africa hands like Juliet and through all the mixed crowd of the specialized safaris devoted to bird-watching, botany, geology, and so on — became clearly explicable to me.

Africa, I now saw, gave something back that Western man had lost. He didn't even need to realize his ancestral loss in order to have anything restored to him. Just to sit in a Land Rover out on the boundless savanna and stare at the wondrous wildlife was enough to bring back into existence his lost sense of belonging to a life-scheme that embraced every living thing in a great, magnetic field of mutual interdependency.

XV

A RARE FINALE

In the final week of our safari it did not seem possible that Kenya and Tanzania could still have more marvels to reveal, especially after the climax at Ngorongoro. As we drove down from the crater highlands into the dry heat of the Rift, I felt so at rest with such familiar phenomena that I said to Lou, "From now on to Nairobi it'll be repetition, thank heaven!" But she continued looking expectantly out the window, refusing to acknowledge my witless remark with even a glance.

Africa, of course, never repeats itself. I should have learned that in the very beginning when I saw my second lion, my second savanna, my second sunset. No two of *anything* in Africa are alike and this lack of repetition is what makes the magic of the African chiaroscuro. Was I unconsciously nearing the saturation point for marvels and resting in the idea of repetition, hoping for the relief of the familiar so that I might review my total African experience through the memory of Gurdjieff's great formulation recalled in Ngorongoro? "The animals are waiting for us to move up. . . ."

Even as I was searching my answers, the rattling Land

Rover was propelling us toward one of the rarest finales ever staged for any safari party that I've heard tell of — a real showstopper followed by an unusual orchid display.

It was to be at Lake Manyara, where we were to see a famous sight, regularly seen only by the park wardens — the tree-climbing lions unique to this national park. And in Tanzania's newest wildlife sanctuary, Tarangire National Park, an ancient baobab had masses of yellow orchids growing out from its crown — an unheard-of symbiosis, even for the experienced Luke.

Manyara is about thirty-five miles from Ngorongoro. Approaching it from the west, we were at first unaware of the grandeur of the high Rift escarpment, where stands the lovely Lake Manyara Hotel with a hundred and twenty-three square miles of national park spread out a thousand feet below. The Tanzania Rift has no opposite wall as in Kenya, where there are cliffs both to the east and to the west. Manyara appeared to be the land's end on first sight — until we looked through the huge telescope mounted on the outside edge of the hotel's view terrace.

As Luke swiveled the telescope for us, the valleylike depression below showed hazy details of trees, a soda lake and one or two elephants or buffalos, which looked like dark boulders until they made a lazy move. I called out the elephants as I spotted them and added, "But no tree lions yet!" because that had been Luke's topic of prophecy on the drive to Manyara. He took my remark seriously and began to explain that one had to be down *on* the valley floor and look *up* to see the tree-climbing lions, which even at close quarters were "devilishly hard to spot," as we would discover next morning. He planned to drive us down with the roof hatch open so Lou and I could look out and *up*. "You'll be my spotters," he said, with a charming smile of anticipation that restrained me from telling him that I really thought those famous tree-climbing lions were overpublicized *if* they existed at all. For who ever heard of a four-hundred-pound lion lying on its backside like a kitten in the treetops? Or, more improbably, slung over tree branches like a sack with head, legs and tail drooping inertly in air?

So when we did at last see a lion sprawled sleeping in a tree, I still could not believe him real. He was contrary to all experience. He was at variance with the laws of nature. He was implausible, unimaginable, unreasonable. Also, he was the most touching sight I had seen in Africa because of his trust in man, for he allowed us to approach from underneath his great stuffed belly only thirty feet above our open hatch and photograph him at will. Indeed, he did not open an eye.

This was our first tree lion and we found him in a magical corner of Lake Manyara National Park, in woodlands of enormous old acacias, their gnarled branches overhead holding thorny canopies — the *Acacia tortilis*, the preferred tree for the lions' siestas aloft. Although Lou and I were riding with heads out the hatch, it was Luke who first spotted the lion through the windshield. Stopping the car, he pointed up and to the left. I could see nothing but a crisscross of branches, of which two or three appeared to be growing downward from a main fork. Those peculiar perpendiculars turned out to be the lion's tufted tail and padded hind feet hanging in space, but I could only see them as erratic tree branches even when I heard Juliet whisper, "*Good* heaven! *Would you believe?*" Not until Luke had backed the Land Rover closer, directly beneath the hanging appendages, did I begin to see the lion.

We had come upon him from the rear. His plump, rufous bottom was seated on a wide limb, his hind legs and tail were hanging free, while his forelegs stretched down to connect with a branch beneath his seat. Against this, his front pads appeared to be pushing to secure a prop for his elevated rear end. When my zoom lens was aimed straight up, I could see a bit of his muzzle — half a nostril and a corner of mouth to show me where his face was, flattened down on the larger branch between his forelegs. The absolute wonder of him, looking as if he had been *flung up* into that tree and had sprawled there as the branch had happened to catch him, filled me with the mad excitement of my first days on safari.

My companions were similarly affected. We all held our breaths and, between camera clicks, the only sounds in the Land Rover were our occasional pantings and the hum of the idling motor. I had the lion in clear focus at about twenty-five feet above the top of the car. If he had let go of his perch, he'd have fallen on our faces.

Luke watched him through binoculars, smiling to himself, perhaps recalling his last year's safari with Juliet when they had made three separate trips down the escarpment in a vain search for these tree-climbing lions. Their presence in the trees was always unpredictable, a mystery like the reason they had climbed them in the first place. The Manyara park booklet gave you a choice of reasons why lions spend so much time resting in trees: "to avoid the unwelcome attention of biting flies;

to catch the breeze; to obtain a view; to find a spot a degree or so cooler than the ground below, or, most probably, to keep out of the way of the buffalo and elephant herds."

When we finished photographing our lion from the rear, Luke backed through the dense thicket around the tree base to find us a possible front view. After a lot of maneuvering he managed to get us in a place beneath the head, which was maneless, therefore female. From this end we now had a new tree lion experience. The giant cat's yellow muzzle was turned sideways on the branch, one paw folded under her chin like a furry pillow. Her eyes were shut in a sleep so deep she hardly seemed to be breathing. The distended belly hanging beneath the branch like an overstuffed sack explained her gluttonous stupor. A recent kill had undoubtedly been an animal as large as a buffalo or zebra, from which she had gnawed off her killer's share of some forty pounds (if you can believe the books!) before yielding the remainder to her mate.

At this point in our safari, we had watched many lions sleeping off their huge feedings after a kill, their bodies stretched out on the ground to their full length of eight or nine feet nose-tip to tail-tip. It was quite another thing now to look *up* at forty pounds of ingested meat hanging overhead in a stretched skin sack that seemed to have no relation to the lesser lion parts surrounding it. How could that lean back and those limp, dangling legs support such a burden of belly? How could they carry such a load up a tree in the first place?

We continued to gaze up in disbelief after we finished photographing. Finally, Juliet put away her camera with a sad headshake. "She's never going to give a sign of life," she said. And right then, without opening an eye, our lioness responded with a sign that we heard before we saw it. A stream of urine splattered down from above, and drenched the spot where we had recently parked to get a rear view of her.

Despite the comic finale of our first encounter with the tree-climbing lions, we all automatically observed our strict rule of silence on game runs. So we waited until we were back on the road before bursting into laughter when Luke said, "Next time you ladies go hunting tree lions, I suggest you wear your rain hats!"

In the late afternoon, in another section of the acacia woodlands, we came to a fork beyond which two safari vehicles were parked at the base of a great tree. Luke indicated the cars to us and said with a grin, "Tree lions anyone?" Of course we all cried "Yes, yes!" although exhausted after a long day of viewing the other wildlife of Lake Manyara's tremendous variety. The spoonbills and elephants had all but done us in.

As we approached the tree, its canopy appeared to be made of the faces of lion cubs peering down through the foliage to see the queer doings of the photographing tourists below them. When we were at last under the tree looking up, we saw their fat little bodies yellow in the sunlight and patterned by the dark lattice from the upper branches where they were clambering awkwardly. Below, sprawled asleep in the larger tree forks, were three or four adult lions whose bodies, like those of the cubs, could be seen only in mosaics through the interstices of the crisscrossing branches.

As for me, our prodigious safari could have ended

The cub looks down on us "with its round, baby-lion eyes"

right there beneath that lion tree, and in one sense it did, although we had three more game parks in our itinerary — Tarangire, Arusha and Amboseli. I was certain that none of them would or could show us that special event, that combination of magical art Manyara had displayed to us who *happened to be there* on that particular day at the sunset hour. Africa's renowned phenomena do not appear regularly on schedule. It's really a matter of luck what you see on safari. For example, there are tourists who have gone to Lake Nakuru without ever seeing the million flamingos around the shorelines, but only a small patch of them in the middle of the lake and these too far away to distinguish the separate birds.

Manyara's treeful of lions held us all silent and bewitched as we sat staring up into the spreading acacia. The cubs, patterned by sunlight and branch shadow, were impossible to count with accuracy. Lou held up nine fingers to show us her guess. I resumed watching them through my viewfinder, sweeping my zoom lens back and forth slowly, waiting for a cub's face to appear. When a face did peer down between branches, followed by an outsized forepaw on which it laid its head so that it might watch us in comfort with its round, baby-lion eyes, I almost fainted with delight. There was no strength left in my arm to hold my camera for a second shot. I dropped back on the seat and gazed up through the hatch at what I believed was my crisis of wonder for a lifetime. I had reached the saturation point. My neck felt permanently damaged from bending backward. Unanswerable questions plagued my wearied mind. How had these tree-climbing lions become so accustomed to the sight and stench of vehicles beneath them that the experienced adults would not even rouse themselves from heavy sleep to spy out below them a possible enemy? How had they transmitted this trustful familiarity to their uneducated cubs?

The sun was making ready to set behind the escarpment when we started to drive back to the hotel. There can be no more wonders, I told myself, I've had them all. An old baobab tree growing out from the Rift wall was a familiar sight by now, a halfway marker between the park and our hotel. It could have nothing new or

212

startling to reveal, for we had passed it often going or coming, although never at this late hour. Casually I looked at it as we passed and now saw it as if for the first time.

The setting sun had called forth the pinkish tones of the baobab's gray bark and was causing it to glow as if from an inner flame. I seemed now to be looking no longer at a rooted member of the floral world but at some animal of flesh with its multiple arms outflung and its whole body twisted in a grotesque contortion, as if trying to communicate its state to anybody who passed by, to anyone who had eyes to see. . . .

On this steep climb there was no possibility of asking Luke to stop that I might photograph the tree and try to discover what it was trying to say. As we ground our way upward, I could only stand on my seat and watch it from the hatch. The side away from the low sun was now drained of its rosy glow. The smooth, heavily folded bark was again gray and ghostlike, as if something in the baobab had died with our unresponsive passing. Now I realized why the baobab had been an object of worship and awe to the primitive African since the most ancient prehistoric times.

It is indeed a ghostly tree, especially when it bears neither leaves nor blossoms, which are due to come forth in October. Seeing it bare-branched as we did, one could understand why David Livingstone had described it in his journal as seeming to have been "planted upside down," thus suggesting an outsized carrot with its roots in the air. As for me, seeing it in a Manyara sunset, the baobab would forever seem a gigantic creature of sorts, with a bombacaceous trunk twenty-five feet in diameter, multiple armlike branches of lesser girth and a final crest of terminal twigs resembling what used to be called "a windblown hair-bob." A tree trying to resemble man — a mutant between the vegetable and animal worlds!

That night on the hotel's wide view terrace, a cocktail party was given by Luke's associates for three of their company's safaris, all of them meeting in Manyara on the homebound stretch. This was a delightful custom of the company, but a bit difficult for me to take after my magical day with lions in trees and the beckoning baobab.

I dressed for the party with inner rebellion. But once out on the terrace, my mood softened when I saw the style of the send-off our escorts had arranged. A long, lamplit table sparkled with bottles of British brand-name liquors and platters of hors d'oeuvres looked like Rumpelmayer creations. The escorts called out to the waiters the names of the preferred drinks of their charges as each appeared. No need to ask, after all the nights of safari sundowners they had shared together. This too, I realized, was a feature of the safari scene that our escorts wanted us to enjoy. And even more restorative than my first highball was the fact that nobody in the other parties attempted to make conversation with us about their day in the park, or expected any from our party. We confined ourselves to a blissful silence during which each, like a miser, sorted over his day's collected impressions. The darkness beyond the edge of the sumptuous terrace looked solid enough to walk on. Beneath it, at the foot of the escarpment, I wondered what the tree-climbing lions were doing now. I would never see those adorable cubs again, except in my pictures, and then only if my shots were successful. But I would see the great old baobab once more, on our way from Manyara to Tarangire. It would be quite another personage in the morning light; perhaps it might even be just a tree. . . .

While packing in our room that night Lou said, "You know that one week from tonight we'll be on the midnight plane for Paris." I had to see it written into our itinerary to believe her. East Africa had claimed me, penetrated me, occupied me as if it were a continuing experience that would never end. Though my sensibilities had long since reached the impossibility of one more impression, another part of me was eagerly asking for more. . . .

One more lion, one more giraffe, one more lilac-breasted roller, one more Masai striding across the savanna on an invisible foot trail narrower than any animal's . . . just one more touch of Africa's sorcery, in any category. Another sunset would do nicely.

Good-bye to Africa? How could one leave this land of magic and miracles, so rugged to traverse, so exhausting in its vastness? I'm like a drug addict, I thought, grabbing for one last shot, one last look, before final separation from this entrancement.

The dry plains country we drove through in the next days was but a duplication of the hundreds of miles of similar grasslands we had crossed before. But now that the awareness of our departure had sharpened my vision, it was a fresh, new loveliness. The enormous, flat vistas of gold grass spread to the gray horizon in all directions.

All was gold and gray, kind to the eyes and, when in the parks, restful to the brain no longer magnetized by the wildlife of the game reserves.

After weeks of reading East Africa's local newspapers, national park magazines and farmers' periodicals, one understood that the key problem about the reserves seemed to be: *How long can they endure?* How long can the game-management people hold off the tribes clamoring for more land to farm, more pasturage for their cattle? With their populations perennially starved for protein food, how could any administration hope to control the poaching in the protected reserves? Even the most indifferent tourist became aware that the rallying cry in these developing young republics was *Africa for the Africans!* Not *Africa for the lions,* or for the elephants, or the rhinos, or for the tourists they attract. Viewed from this perspective, it was not understandable that so many of the large game reserves set up by the British had survived not only the first lawless impulses of Independence but nearly a decade of black rule. Africa's own scientists, university trained in land- and game-management, themselves often asked (between the lines of their published pieces on the problems of wildlife conservation) *How long?* as if they suspected that the endangered species they were striving to protect might very well turn out to be themselves.

No safari escort or accompanying game warden, black or white, would be drawn into any discussion about the problems or politics of the countries we were driving through. I had learned this on our first days of safari when, in some hotel, I had found a week-old newspaper article about some kind of shooting trouble going on between Tanzania and Uganda along the northwest border. Luke only smiled at our anxious queries. "Nothing to worry about," he said, then lapsed into a scowling silence.

Now in the windup of our safari, Tarangire in Tanzania was interesting to explore because it showed us what all the game reserves must have been like at first, after big-game hunting was strictly prohibited. Made a reserve in June 1970, only some fifteen months before our arrival, Tarangire's previous use as an authorized area for licensed professional hunters and their clients was well remembered by the terror we found in its wildlife. This was the only reserve we had visited where lions showing fear of man walked away growling when our Land Rover appeared on the scene. After the illusions of "peaceable kingdoms" shared by man and beast, which the older parks like Serengeti, Ngorongoro and Manyara fostered, Tarangire's wary wildlife, remembering the guns of the hunters, made our game-viewing here a sad experience. How many lion generations must pass before man would be accepted here with the touching trust of those Manyara lions? However long this might take, you knew that it must happen eventually, thanks to — of all things! — Tarangire's tsetse fly infestation. This fly was deadly to domestic cattle and thus it protected the park from tribal demands to use it as grazing lands! The wild animals had built up an immunity to sleeping sickness, and human beings were supposedly not affected by this particular form of trypanosomiasis. Nevertheless, we made our run through the "tryps" zone (fortunately limited) armed with tough rubber flyswatters and cans of insecticidal spray.

At Tarangire our tents are guarded by three baobab trees

The only pictures I took in Tarangire were of the bao-babs. This was real "baobab country," a dry wilderness where the tall trees stood forth like gray guardsmen spaced out across the stony landscape. Three of them towered over our tents in much closer relation to us than Manyara's baobabs had been. We could run our hands over the fleshy folds of their trunks. We could look up through their bare, distorted branches and see their curi-ous seedpods hanging on high. A single blossom had produced each one. It was a waxy, white flower, Juliet said, five-petaled and pendulous, which lasted for only one day. Every detail I learned about this wondrous tree increased my amazement. Since every part of the baobab, from bark to pulp to leaves and fruit, was edible to either elephant or man, how could there be so many of them still standing unscarred, waving their arms, it seemed, as if to each other across the windy plains of Tarangire. There was one standing alone in a stony field just outside the lodge grounds, where it waved as if for our attention as we drove away. "Stop just a minute!" cried Juliet and pointed to the tree.

We all stared. The baobab looked like a crazy old beggar, naked except for sprays of tiny yellow orchids sprouting out like flowery hair from its topmost branches, a character out of Marat-Sade. Juliet, looking through her glasses at an orchid, said it must be one of the *Dendrobium* species, possibly the *Ansellia nilotica* or "Leopard Orchid." We must take pictures back home with us for her positive identification.

"How could we have missed this extraordinary sight all the other times we drove by it?" Juliet exclaimed.

Amboseli Game Reserve in Kenya was our last stop before Nairobi, a hundred and fifty miles to the north, our last chance to look at a free-living lion, or elephant, or rhino; our last night in the bush under canvas in the rustic luxury (every tent with shower and toilet) of Amboseli Safari Camp. The first thing one sees in Amboseli is the snow-covered cone of Kilimanjaro floating high in the sky across some fifty miles of the absolutely flat plain that stretches all the way into Tanzania, where the legendary mountain rises. Having just come from Tanzania without having had a glimpse of Kilimanjaro (due to cloud cover, intervening foothills, etc.) the first sight of it is one of Africa's most dramatic experiences. You are looking at it actually from another country. This, the

219

fourth highest mountain in the world, does not seem to be based on earth. Rather, it is a fixture of the African sky, rootless and vaporous above the burning plain.

Like Tarangire, but for opposite reasons, Amboseli showed what unrestricted tourism over a long period, coupled to unregulated running of cattle over the reserve by the Masai, can do to impoverish an ecosystem that must once have been in magnificent balance.

Now, at the end of the dry season, Amboseli looked like a dust bowl, except in the central swamp area — the wildlife sanctum where the watering holes are. These dusty surroundings were crisscrossed by tire tracks made by unrestrained tourists chasing big game across the plains, or provoking it to chase them. These graffiti of thoughtless travelers who would tear up the last shreds of a dying ground cover for the sake of a little "fun" caused me to choke with anger, as well as dust. I heard myself asking *How long?* and ached with guilt as I watched some of the reserve blowing away behind our own wheels, even though Luke kept scrupulously to the authorized track.

Very shortly, as always seemed to happen when one "worried" about Africa, this continent gave us a resounding proof of the imperishable nature of a world without end. Around a curve in the track, we came upon the largest pride of lions we had ever encountered. They lay in the speckled shade of a dead thornbush and a dying fever tree. We counted twelve as Luke crept nearer — males, females and cubs — and from where he stopped on a slight elevation we saw there were four more. This, you knew at once, was no wandering group of displaced lions going nowhere. It was an established lion community, adjusted to a life in the dust bowl on the fringe of the sanctuary swamp and intending to remain there within easy distance of their self-replenishing commissary — the watering holes to which every animal of the plains, including the Masai cattle, had to come to drink. The lions' glossy coats, slightly grayed with the dust in which they lay, showed no bony protuberances or hints of rib-cage shadows.

Lou and Juliet took pictures in the hazy late-afternoon light, but I simply sat and gazed at the tremendous pride sprawled out before us with pictorial abandon. This I knew was the last wildlife impression I would take out of Africa. It was the last I could truly register, with mind and emotions.

The lions that had arrived there first and captured the deepest shade were stretched out along the length of the heaviest tree-trunk shadow; in the narrower shadows of the branches slanting into it, other elongated bodies formed a tree of life on the ground. It took only three adults to design the trunk of this tree and sometimes only one could make a ten-foot branch.

Any game warden would have laughed at my interpretation of that tree-shaped cryptogram, but I chose to read it as the final stunning affirmation of the endurance of African life. My wonderful safari ended right there for me. All else was but a postlude — a final night in the bush, a final drive up the highway to Nairobi, and a last night in fabulous Africa before flying to lovely, tame Paris.

Kilimanjaro was the real end for us now. It had disappeared behind heavy cloud cover when we got back to camp and you couldn't recall how high in the sky you had to search in hope of detecting a trace of its icy shadow. Without that stupendous backdrop, Amboseli Safari Camp was like any stretch of Arizona desert minus the cactus. Until, of course, after sunset, when you could listen to the sounds. Then you knew well you were still in Africa.

We began listening to them in the bar-restaurant tent where we had our last safari sundowners together, before dinner. I heard the birds going to bed as Luke raised his glass more than once to his ladies. He told us we had been the easiest trio he had ever taken out but I knew that he lied in his fine white teeth. By my infallible measuring eye, he was at least five pounds lighter than when he had picked us up one month and over two thousand miles ago, in the Nairobi airport. As he ordered again all around, the avian chorus slowly came to its sleepy end.

Later, on my cot in tent number 10, I listened for the last time to the sounds of the bush, the sounds of Africa. . . . Could they not be arranged into a sort of "symphony safari" for the blind? The sunset prelude, the increase in sound volume as the darkness comes on, the rising in a crescendo of hoots, moans, mutterings and howls, and then a final coda that might be the single death shriek, or cough, or groan of an animal brought to ground by the predator. As I lay there with eyes closed, "seeing" Africa through its sounds, I heard its leaping life going on all around me out in the great darkness of Amboseli's deceptively lifeless bush.

A baboon screamed in a hair-raising voice of a man being murdered. The dreadful sound went on for several seconds, then stopped abruptly mid-scream. I guessed that a lion or leopard, the only carnivores strong enough to take a fierce-fighting baboon, had got him. This peculiar awareness of the events in an African night did not seem in the least strange to me. Anyone who had entered into the safari scene as I had, with love and total attention, would have developed it, like a sixth sense. I wondered only, like a possessive member of his family, what the devil my baboon brother was doing down on the ground, out of the safety of his sleeping tree at this time of night!